Lecture Notes in Mathematics

Edited by A. Dold and B. Eckmann

522

Clifford O. Bloom
Nicholas D. Kazarinoff

Short Wave Radiation Problems in Inhomogeneous Media: Asymptotic Solutions

Springer-Verlag
Berlin · Heidelberg · New York 1976

Authors
Clifford O. Bloom
Nicholas D. Kazarinoff
Department of Mathematics
State University of New York
at Buffalo
Amherst, N. Y. 14226/USA

Library of Congress Cataloging in Publication Data

Bloom, Clifford O 1935-
The asymptotic solution of high-frequency radiation-scattering problems in inhomogeneous media.

(Lecture notes in mathematics ; 522)
Includes index.
1. Radiation. 2. Scattering (Physics) 3. Asymptotic expansions. I. Kazarinoff, Nicholas D., joint author. II. Title: The asymptotic solution of high-frequency radiation-scattering problems ... III. Series: Lecture notes in mathematics (Berlin) ; 522.
QA3.L28 no. 522 [QC475] 510'.8s [539'.2] 76-17818

AMS Subject Classifications (1970): 35B40, 35B45, 35J05, 53C25, 78A05, 78A40

ISBN 3-540-07698-0 Springer-Verlag Berlin · Heidelberg · New York
ISBN 0-387-07698-0 Springer-Verlag New York · Heidelberg · Berlin

This work is subject to copyright. All rights are reserved, whether the whole or part of the material is concerned, specifically those of translation, reprinting, re-use of illustrations, broadcasting, reproduction by photocopying machine or similar means, and storage in data banks.

Under § 54 of the German Copyright Law where copies are made for other than private use, a fee is payable to the publisher, the amount of the fee to be determined by agreement with the publisher.

© by Springer-Verlag Berlin · Heidelberg 1976
Printed in Germany

Printing and binding: Beltz Offsetdruck, Hemsbach/Bergstr.

PREFACE

These notes are based upon a series of lectures given at the University of Oxford, Spring, 1975 by the second author. The authors thank Dr. J. B. McLeod for the opportunity for their joint work to be presented in his seminar.

These notes are primarily concerned with existence, uniqueness, a priori estimates, and the rigorous asymptotic solution of the radiation-scattering problem:

$$\text{(P)} \qquad \begin{aligned} \Delta u + \lambda^2 n(x)u &= f(x) \qquad (x \in V), \\ u &= u_0(x) \qquad (x \in \partial V), \\ \lim_{R\to\infty} \int_{r=R} r\left|u_r - i\lambda u + \frac{m-1}{2r}u\right|^2 dS &= 0 \quad (|x| = r) \end{aligned}$$

for λ large. Here V is the exterior of a not necessarily convex or star-shaped body $\partial V \subset R_m$ $(m = 2 \text{ or } 3)$. In Chapter 1 we obtain new point-wise and L_2 a priori estimates by a variation of K. Friedrichs' abc-method. These estimates imply uniqueness of the solution u of the boundary-value problem (P) above. We construct an approximate solution (in powers of λ^{-1}) to the problem (P) in Chapter 3. We also construct there an approximate solution to the more general radiation-scattering problem where the values of a linear combination of u and its normal derivative are prescribed on ∂V. We apply the a priori point-wise estimate of Chapter 1 to prove in Chapter 3 that the approximate solution to the problem (P) is an asymptotic expansion of the exact solution as $\lambda \to \infty$. This asymptotic approximation yields a high-frequency asymptotic expansion of the leading term of the far field scattering amplitude $a(x,\lambda)$, where

$$a(x,\lambda) = \lim_{r\to\infty} (r e^{-i\lambda r} u).$$

In Chapter 2 we study the reciprocal relationship between ray systems in the inhomogeneous medium with index of refraction $n^{\frac{1}{2}}$ and the smoothness and asymptotic properties of $n^{\frac{1}{2}}$ as $r \to \infty$. This study, which is the first of its kind insofar as we know, is necessary to give meaning to the computations involved in determining the formal approximate solution of problem (P). It is also a necessary ingredient in using the a priori estimates from Chapter 1 to show in Chapter 3 that truncations of the formal asymptotic series uniformly approximate the solution u in the closure of V. The results of Chapter 2 are obtained through application of the fixed point theorem for contractions to the integral equation form of the ray equations. In Chapter 4 we show how results of D. Eĭdus imply existence of u; and we also give an alternative existence proof, based on his work, which may be susceptible of generalization to elliptic equations of the form

$$\nabla \cdot (E(x)\nabla u) + \lambda^2 n(x)u = f(x),$$

in cases where $E(x)$ and $n(x)$ are not constant outside a compact set.

The authors thank Professor J. B. Keller for his encouragement and helpful advice.

Amherst, New York
December, 1975

CONTENTS

CHAPTER 0

INTRODUCTION

Much effort has been devoted to studying the solutions of radiation-scattering problems of the form

(S)

(0.1) $$\nabla \cdot (E(x)\nabla u) + \lambda^2 n(x)u = f(x) \qquad (x \in V)$$

(0.2) $$\alpha(x,\lambda)u + \beta(x)\nu^*(x) \cdot \nabla u = g(x)$$

$$(x \in \partial V,\ \nu^* = \text{unit exterior normal to } \partial V)$$

(0.3) $$\lim_{R\to\infty} \int_{r=R} r\left|\frac{\partial u}{\partial r} - i\lambda u + \frac{(m-1)}{2r}u\right|^2 dS = 0$$

$$(m = 2,3\ ;\ r = |x| = [\Sigma_1^m (x^i)^2]^{\frac{1}{2}})\ ,$$

where V is the exterior of one or more scattering obstacles of finite cross-section, $E(x)$ is a strictly positive definite matrix that tends to the identity matrix as $r \to \infty$, $n(x)$ is a strictly positive function that tends to 1 as $r \to \infty$, and λ is real and positive. The inhomogeneous term in (0.1) is usually assumed to vanish outside a compact set, or to approach zero at a prescribed rate as $r \to \infty$. A special case of interest is $f(x) = \delta(x,x_0)$ and $g(x) = 0$. Then the solution of Problem (S) is a Green's function.

The interest in the scattering Problem (S) arises because the equations (0.1) - (0.3) are a mathematical model for the propagation of time harmonic waves in an inhomogeneous medium filling the exterior region V. If $u(x,\lambda)$ is a solution of (S) with $\alpha(x,\lambda) = \alpha_1(x) - i\lambda c\,\alpha_2(x)$, then $u(x,\lambda)e^{-i\lambda ct}$ is a time harmonic solution of frequency $\omega = c\lambda$ of the wave equation

(0.4) $$\nabla \cdot (E(x)\nabla W) - \frac{n(x)}{c^2}W_{tt} = f(x)e^{-ic\lambda t} \qquad ((x,t) \in V \times (0,\infty))$$

$$(c = \text{speed of propagation of signals if } n(x) \equiv 1)$$

that satisfies the boundary condition

(0.5) $$\alpha_1(x)W + \alpha_2(x)W_t + \beta(x)\nu^*(x) \cdot \nabla W = g(x)e^{-ic\lambda t}$$

$$((x,t) \in \partial V \times (0,\infty))\ ,$$

and the radiation condition

(0.6) $$\lim_{R\to\infty} \int_{r=R} r|\mathcal{D}_1 W|^2 dS = 0 \qquad (t > 0)\ ,$$

where

$$\mathcal{D}_1 W = \frac{\partial W}{\partial r} + [-i\lambda + \frac{(m-1)}{2r}]W \qquad (m = 2,3)\ .$$

The time harmonic solution $ue^{-ic\lambda t}$ of (0.4) - (0.6) has been shown (under certain conditions on the coefficients $E(x)$, $n(x)$, $\alpha_1(x)$, $\alpha_2(x)$, $\beta(x)$, and the source terms $f(x)$ and $g(x)$) to be the steady state solution of the following initial-boundary value problem.

$$(0.7) \qquad \nabla \cdot (E(x)\nabla W) - \frac{n(x)}{c^2} W_{tt} \equiv f(x)e^{-i\lambda ct} \qquad ((x,t) \in V \times [0,\infty)) ,$$

$$(S') \qquad (0.8) \qquad \alpha_1(x)W + \alpha_2(x)W_t + \beta(x)\nu^*(x) \cdot \nabla W = g(x)e^{-i\lambda ct} \qquad ((x,t) \in \partial V \times (0,\infty)) ,$$

$$(0.9) \qquad W(x,0) = h_1(x) , \quad W_t(x,0) = h_2(x) \qquad (x \in V) .$$

If $f(x)$, $h_1(x)$ and $h_2(x)$, $n(x) - 1$ and $E(x) - I$ have compact support (and are sufficiently smooth), then scattering theory [13, p. 164] can be applied to show that the solution $W(x,t)$ of (S') approaches $ue^{-ic\lambda t}$ as $t \to \infty$ at every point of V . For example, D. Eĭdus [5,6] has proved such a result if $\alpha_2(x) = 0$, $\beta(x) = 0$, $g(x) = 0$, V is an open exterior region with a finite boundary and $\mathrm{supp}(E - I)$ is compact.

In the case of scattering by a single obstacle of finite cross-section that does not "trap" rays (see [13, p. 155]) it is reasonable to expect that the solution W of (S') should approach a time harmonic steady state if $E(x)$ and $n(x)$ are sufficiently smooth. C. S. Morawetz [15] and R. Buchal [4] using different arguments have shown, for solutions of the wave equation defined outside a star-shaped obstacle in $\mathbb{R}^3$, and satisfying a Dirichlet boundary condition that

$$W(x,t) = u(x,\lambda)e^{-i\lambda ct} + \mathcal{O}(t^{-1}) \qquad (x \in V) \quad \text{as} \quad t \to \infty .$$

C. O. Bloom [3] has established an algebraic rate of approach to the steady state for solutions of (S') , defined outside a star-shaped body if $h_1(x)$ and $h_2(x)$ have compact support,

$$\int_V r|f|^2 \, dV < \infty , \quad \alpha_2(x) = \beta(x) \equiv 0 , \quad g(x) = 0$$

and provided $E(x) - I$, $n(x) - 1$ are sufficiently smooth (lie in $C^1(\overline{V})$, $\overline{V} = V \cup \partial V$), $E(x) - I = \mathcal{O}(r^{-1-\delta})$ $(0 < \delta < 1)$, and $n(x) - 1 = \mathcal{O}(r^{-1-\delta})$ as $r \to \infty$.

The argument of Bloom assumes the existence of a unique solution of (S) under these conditions on $E(x)$ and $n(x)$. As far as we know, such an existence-uniqueness result has not been established. We prove the existence of a unique solution of (S) in Chapter 4 of these notes if $\beta(x) \equiv 0$, $\alpha(x,\lambda) \equiv 1$, and $E(x) \equiv I$. In our proof $n(x) - 1$ is not required to have compact support, but it is assumed to satisfy the hypotheses of Theorems 7.1 and 8.1 of Chapter 1. The a priori estimates established in these theorems immediately yield uniqueness of solutions to (S) in this case.

Our existence theorem is alternative to a theorem of D. M. Eĭdus [6]. As is often the case with a priori bounds, if they imply uniqueness, then they imply existence as well. The idea of our proof is to consider a sequence of modified problems (S) with n replaced by n_j , where $\mathrm{supp}(n_j - 1)$ is compact and expands to all of V as

$j \to \infty$, and to show that solutions to these problems converge to a limit that is the desired solution of (S) when $E(x) \equiv I$, $\beta(x) = 0$ and $\alpha(x,\lambda) = 1$. Existence of solutions u_j to the modified problems is implied by Eĭdus' result. Applying the a priori estimates derived in Chapter 1 of these notes allows us to conclude that these solutions form a Cauchy sequence, which converges to $u(x,\lambda)$.

If $E(x) \equiv I$, equations (0.1) - (0.3) may govern the propagation of electromagnetic waves of frequency $\omega = \lambda c$ in an optical medium, where the index of refraction is $n^{\frac{1}{2}}(x)$. The solution of (0.4) - (0.6) is the amplitude of the scalar potential of the time-harmonic electromagnetic field. Under certain conditions on the shape and the physical properties of ∂V , the components of the electric field intensity and the magnetic field intensity also satisfy (0.4) - (0.6).

If $E(x) = n(x)I$, equations (0.1) - (0.3) govern the propagation of acoustic waves in a slightly compressible medium of density $\rho(x) \equiv 1/cn(x)$; see [8, Chapt. 1]. Under the acoustical interpretation the solution of (0.4) - (0.6) is the excess pressure or the velocity potential of the time harmonic field. In many important applications, the wave length $\lambda \gg 1/a$, where a is the minimum diameter of the scattering obstacle.

In Chapter 1 of these notes we obtain a priori estimates for the solution $u(x,\lambda)$ of the following radiation-scattering Problem:

(P)

$$(0.10) \qquad Lu = f(x,\lambda) \qquad (x \in V \subset \mathbb{R}^m ; m = 2,3 ; \lambda > 0) ,$$

$$(0.11) \qquad u = g(x) = u_0(x) \qquad (x \in \partial V) ,$$

$$(0.12) \qquad \lim_{R\to\infty} \int_{r=R} r|\mathcal{D}_1 u|^2 = 0 ,$$

where

$$Lu = \Delta u + \lambda^2 n(x) u .$$

If $m = 2(3)$, then V is the exterior of a smooth closed curve (surface) ∂V (a smoothly embedded $(m-1)$ sphere in $\mathbb{R}^m$). We assume that ∂V can be illuminated by convex surface (curve) contained in V (see Definition 2.1 of Chapter 1).

We require that

(H)

(i) $u_0(x) \in C^1(\partial V)$,

(ii) $n(x) \in C^2(\overline{V})$,

(iii) $f(x,\lambda) \in C(\overline{V})$ <u>for every</u> $\lambda > 0$,

(iv) $\int_V r^2|f|^2 < \infty$, and

(v) $n(x) \geq n_0 > 0$ <u>for all</u> $x \in \overline{V}$. In addition we require that <u>as</u> $r \to \infty$

(vi) $|n(x) - 1| = \mathcal{O}(r^{-p})$ <u>for some</u> $p > 2$,

(vii) $\nabla n(x) = \mathcal{O}(r^{-2})$,

(viii) $\partial^{i+j} n(x)/\partial x^i \partial x^j = \mathcal{O}(r^{-3})$ $(i+j = 2 , \quad i,j \geq 1)$.

Here $\overline{V}$ is the closure of V .

Most of Chapter 1 is devoted to obtaining estimates for the "energy norms"

$$\|\nu^* \cdot \nabla u\|_{\partial V} = \left(\int_{\partial V} |\nu^* \cdot \nabla u|^2\right)^{\frac{1}{2}}$$

(ν^* = exterior unit normal to ∂V),

$$\|r^{-1}\nabla u\|_V = \left(\int_V r^{-2}|\nabla u|^2\right)^{\frac{1}{2}} ,$$

and

$$\|r^{-1}u\|_V = \left(\int_V r^{-2}|u|^2\right)^{\frac{1}{2}} .$$

We find that as $\lambda \to \infty$ (we let $a=1$ for convenience)

$$\|\nu^* \cdot \nabla u\|_{\partial V} , \ \|r^{-1}\nabla u\|_V \leq \Gamma_1 N(f,u_0 ; \lambda) , \tag{0.13}$$

and

$$\|r^{-1}u\|_V \leq \Gamma_2 \lambda^{-1} N(f,u_0 ; \lambda) , \tag{0.14}$$

where Γ_1 and Γ_2 are constants that depend only on ∂V and $n(x)$, and

$$N(f,u_0 ; \lambda) = [\lambda \operatorname*{Max}_{\partial V} |u_0| + \|u_{0T^*}\|_{\partial V} + \|rf\|_V] .$$

We use (0.13) and (0.14) to derive an upper bound for the field strength $|u(x,\lambda)|$ that holds uniformly on $\overline{V}$ as $\lambda \to \infty$:

$$|u(x,\lambda)| \leq \Gamma_3 \lambda^{(1+m)/2} r^{(1-m)/2} N(f,u_0 ; \lambda) , \tag{0.15}$$

where Γ_3 is constant that depends only on ∂V and $n(x)$.

The estimates we obtain for the L_2 norms of u/r and $\nabla u/r$ also imply an upper bound on the energy $E_{R_0}(ue^{-i\lambda ct})$ of the function $ue^{-i\lambda ct}$ that is contained in the region $V(R_0)$ between the boundary ∂V of the scattering obstacle and a sphere of radius R_0 ; namely,

$$E_{R_0}(ue^{-i\lambda ct}) \leq \Gamma_3 N(f,u_0 ; \lambda) , \tag{0.16}$$

where $\Gamma_3 = 2(\operatorname*{Min}_{V(R_0)} r^{-2}) [\Gamma_1^2 + \Gamma_2^2]$. As we mentioned above, these same estimates immediately imply uniqueness of u .

In Chapters 2 and 3 we consider the following Problem (U) :

<u>Let</u> $u(x,\lambda)$ <u>be the solution of equation</u> (0.10) <u>subject to the radiation condition</u> (0.12), <u>and the boundary condition</u>

$$\alpha(x,\lambda)u + \beta(x,\lambda)\nu^*(x) \cdot \nabla u = g(x,\lambda) \qquad (x \in \partial V) . \tag{0.17}$$

<u>Construct an asymptotic approximation</u> $u_M(x,\lambda)$ <u>of</u> $u(x,\lambda)$ <u>such that</u>

$$u(x,\lambda) - u_M(x,\lambda) = \mathcal{O}\left(\lambda^{-M+\frac{1}{2}(1+m)} r^{-\frac{(m-1)}{2}}\right) \qquad (M > \tfrac{1}{2}(1+m)) \tag{0.18}$$

<u>uniformly in</u> x $(x \in \overline{V})$.

We use the notation $\overline{S}$ for the closure of a set S . We call this problem the Ursell radiating body problem; see F. Ursell [19].

We apply the a priori estimate (0.15) to solve problem (U) for a general class of scattering obstacles in the case $\beta \equiv 0$, $\alpha(x,\lambda) \equiv 1$, under physically reasonable hypotheses on (i) the smoothness of ∂V, $g(=u_0)$ and f, and (ii) the asymptotic behavior as $r \to \infty$ of f, n and derivatives of these functions. We assume for simplicity that $f(x,\lambda) = f_0(x)$ and $g(x) = u_0(x)$, where $f_0(x)$ and $u_0(x)$ are independent of λ. The asymptotic approximations we obtain satisfy (0.18) for positive integer values of M. Note also that $\lim_{r\to\infty} re^{-i\lambda r}u_M$ is an asymptotic expansion of the scattering amplitude of u. The function $u_M(x,\lambda)$ is constructed to satisfy the radiation condition (0.12), the boundary condition (0.11), and to have the property that as $\lambda \to \infty$

$$(0.19) \qquad L u_M(x,\lambda) = f_0(x) + \mathcal{O}\left(\lambda^{-M} r^{-\frac{(m+3)}{2}}\right) \qquad (x \in \overline{V}; m=2,3) .$$

To get (0.18) we apply the point-wise estimate (0.15) to $u - u_M$. Under conditions similar to those conditions that we impose on ∂V, $n(x)$, g and f to construct $u_M(x,\lambda)$ in the case $\alpha \equiv 1$, $\beta \equiv 0$ our method can be applied to yield a function $u_M^*(x,\lambda)$ that satisfies (0.10), (0.12) the boundary condition (0.17), and which also has the property that

$$(0.20) \qquad L u_M^*(x,\lambda) = f(x,\lambda) + \mathcal{O}\left(\lambda^{-M} r^{-\frac{(m+3)}{2}}\right) \qquad (x \in \overline{V}; m=2,3) .$$

We describe the procedure in Chapter 3 assuming that (i) $\alpha(x,\lambda)$, $\beta(x,\lambda)$ are sufficiently smooth in x, (ii) $\alpha(x,\lambda) = \alpha_1(x,\lambda) - i\lambda\alpha_2(x)$, (iii) $D^p f(x,\lambda)$, $D^p g(x,\lambda)$, $D^p\alpha_1(x,\lambda)$, $D^p\alpha_2(x)$, $D^p\beta(x,\lambda) = \mathcal{O}(1)$ as $\lambda \to \infty$, $|p| = 0,1,2,3,\ldots$, (iv) $\alpha_1(x,\lambda)$, $\alpha_2(x)$, $\beta(x,\lambda)$ are of constant sign on the subset of ∂V contained in $\operatorname{supp} g \cup \operatorname{supp} f$; see Chapter 3, Section 6 of these notes. But unless $\beta \equiv 0$, we have no a priori estimate available that can be used to prove that the function $u_M^*(x,\lambda)$, which is an approximate solution of the boundary value problem (0.10), (0.12) and (0.17), is also an asymptotic approximation of $u(x,\lambda)$ as $\lambda \to \infty$ in the sense that (0.18) holds.

In the case $n(x) \equiv 1$ we require that the subset of ∂V contained in the support of the radiating sources $f_0(x)$ and $g(x)$ consist of a finite number of disjoint, locally convex "patches" S_i $(i=1,2,\ldots,K)$ joined together so that ∂V is smooth; see Fig. 0.1. In addition, we impose the condition that (i) each straight line ray (since $n(x) \equiv 1$) emanating orthogonally from the patch S_i extends to infinity without again meeting ∂V. If all of ∂V is contained in the support of f or g, then the above requirements are satisfied if and only if ∂V is convex; see Fig. 0.2. In order to apply the a priori estimate (0.15) to $u - u_M$ we also need to postulate that ∂V can be illuminated from the exterior.

In the case $n(x) \not\equiv 1$ we impose analogous restrictions on ∂V. The portion of ∂V contained in the support of f or g should consist of a finite number of disjoint patches S_i that are "locally convex relative to $n^{\frac{1}{2}}(x)$", and joined together to form a smooth surface (curve) in $\mathbb{R}^3$ $(\mathbb{R}^2)$. For a patch S_i to be

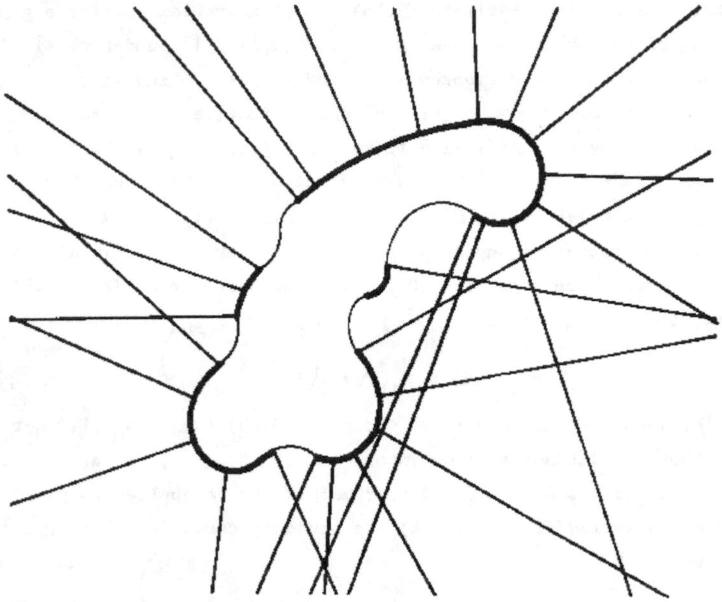

Figure 0.1

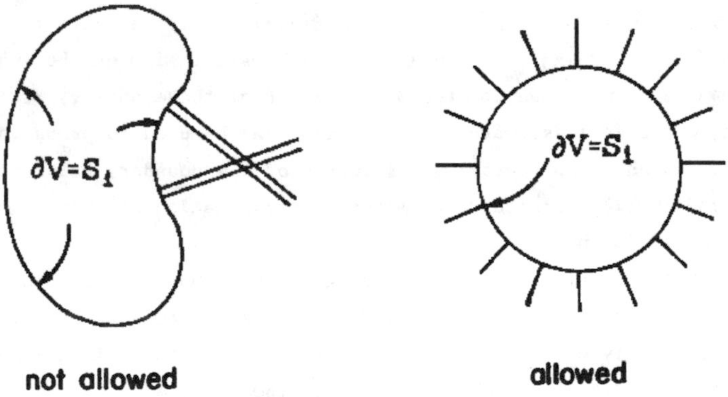

Figure 0.2

locally convex relative to $n^{\frac{1}{2}}(x)$ we first require that each geodesic of the Riemannian metric $ds = n^{\frac{1}{2}}|dx|$, emanating normally from S_i extends to infinity without intersecting itself or again meeting S_i. Second, we require that each member of every pair of geodesics (rays) emanating normally from S_i extends to infinity without intersecting the other. The normal congruence of rays in the tube $T_i \subset \overline{V}$ bounded by S_i and the geodesics from the boundary of S_i form what is called a <u>field</u> $\mathfrak{F}_1$ on $\overline{T}_i$; see [14, p. 108]. The rays and the system of "wave fronts" Σ_i orthogonal to them define a simple covering (a coordinate system) on $T_i(\subset\overline{V})$; see Fig. 0.3. To preclude multiple reflections we impose the additional requirement that no ray emanating orthogonally from S_i intersect $\partial V - S_i$.

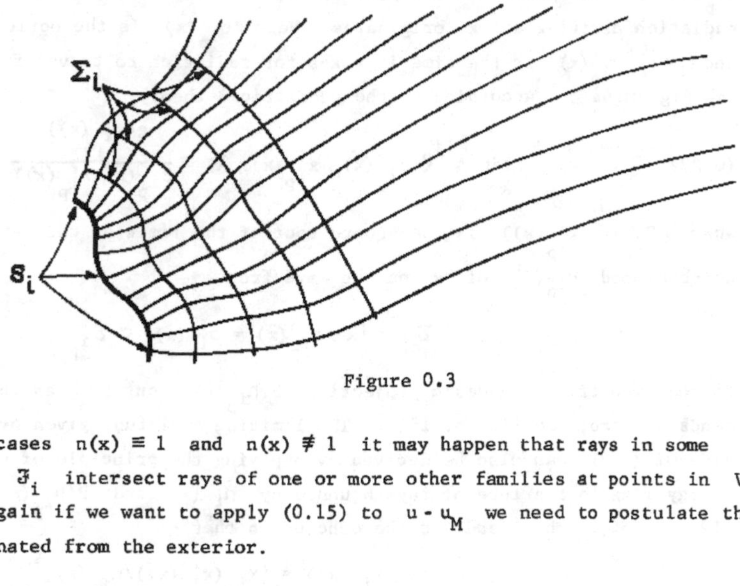

Figure 0.3

In both the cases $n(x) \equiv 1$ and $n(x) \not\equiv 1$ it may happen that rays in some particular family $\mathfrak{F}_i$ intersect rays of one or more other families at points in V ; see Fig. 0.4a. Again if we want to apply (0.15) to $u - u_M$ we need to postulate that ∂V can be illuminated from the exterior.

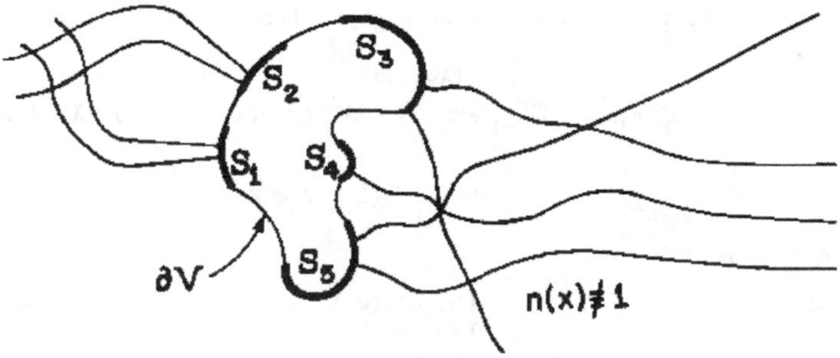

Figure 0.4a

The solution of problem (P) can be written as the sum of $U^1(x,\lambda)$ and $U^2(x,\lambda)$, where $U^1(x,\lambda)$ is the solution of (P) if $f \equiv 0$, and $U^2(x,\lambda)$ is the solution of (P) if $g \equiv 0$. According to the classical geometrical theory of wave propagation, which ignores diffraction, if $V_i \cap V_j = 0$ $(i \neq j)$, then $U^1(x,\lambda)$ is approximately equal to

$$U_N^1(x,\lambda) = \Sigma_{p=1}^{P} e^{i\lambda\sigma_{i_p}(x)} A^{1p}(\sigma_{i_p}(x), x'_{i_p}(x); \lambda) \qquad (1 \leq P \leq K) \tag{0.21}$$

if $\lambda \gg 1$ and $x \in \bar{V}$. Here $x'_{i_p}(x)$ $(p = 1,2,\ldots,P; 1 \leq P \leq K)$ is the point on the patch S_{i_p} that can be joined by an arc A_{i_p} of a smooth geodesic in $\mathfrak{F}_{i_p}$. The arc A_{i_p} is the optical path of least time from x to any point on S_{i_p}. The $x_{i_p}(x)$ are the points on the patches S_i $(i = 1,2,\ldots,K)$ from which most of the radiation arriving at x originates. Here $\sigma_{i_p}(x)$ is the optical length of A_{i_p}, and $t = c^{-1}\sigma_{i_p}(x)$ is the time it takes for radiation to travel from $x'_{i_p}(x)$ to x; see Fig 0.4a. According to the geometrical theory,

$$\lim_{\lambda\to\infty} A^{1i_p}(\sigma_{i_p}(x), x'_{i_p}(x); \lambda) \equiv \frac{g(x'_{i_p}(x))}{C_{i_p}(x, x'_{i_p}(x))} \tag{0.22}$$

where $C_{i_p}(x, x'_{i_p}(x))$ is the square root of the limit of the ratio of the area of a neighborhood $\mathfrak{n}_{i_p}(x)$ of x on the wave front

$$\Sigma_{ct} = \{\hat{x} : \sigma_{i_p}(\hat{x}) = \sigma_{i_p}(x)\} \subset \Sigma_{i_p}$$

to the area of its geodesic projection $P(\mathfrak{n}_{i_p}(x))$ on S_{i_p} as the area of $\mathfrak{n}_{i_p}(x)$ tends to zero; see [12, p. 154]. The limiting amplitude given by the right hand side of (0.21) can also be derived by applying the principle of conservation of energy flux to the tube of rays bounded by $\mathfrak{n}_{i_p}(x)$ and $P(\mathfrak{n}_{i_p}(x))$ as described in [12, p. 154]. This leads to the conclusion that

$$C_{i_p}(x, x'_{i_p}(x)) = [K_{i_p}(x'_{i_p}(x))/K_{i_p}(x)]^{\frac{1}{2}},$$

where $K_i(x)$ is the expansion coefficient of the wave front $ct = \sigma_i(x)$.

Similarly, the geometrical theory predicts that $U^2(x,\lambda)$ is approximately equal to

$$U_N^2(x,\lambda) = \Sigma_{p=1}^{P} e^{i\lambda\sigma_{i_p}(x)} A^{2p}(\sigma_{i_p}(x), x'_{i_p}(x); \lambda) + B(x,\lambda), \tag{0.23}$$

$$(1 \leq P \leq K)$$

if $\lambda \gg 1$, that

$$\lim_{\lambda\to\infty} \lambda^2 B(x,\lambda) = f(x)/n(x) \qquad (x \in \bar{V}), \tag{0.24}$$

and that

$$\lim_{\lambda\to\infty} \lambda^2 A^{2p}(\sigma_{i_p}(x), x'_{i_p}(x); \lambda) = \frac{-f(x'_{i_p}(x))}{n(x'_{i_p}(x))C_{i_p}(x, x'_{i_p}(x))} \quad (0.25)$$

In the special case that all of ∂V is convex relative to $n^{\frac{1}{2}}(x)$, there is a unique optical path A_1 of least time from x to $x'_1(x)$ on $S_1 = \partial V$, $K=1$ in (0.21) and (0.23), and the summations in (0.21) and (0.23) reduce to a single term; see Fig. 0.4b.

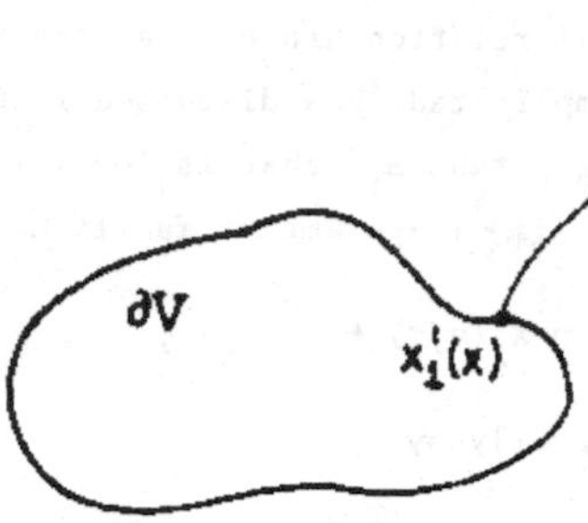

Figure 0.4b

In the case $n(x) \equiv 1$, there is only the problem of choosing the amplitude functions $A^{ip}(\sigma_{i_p}(x), x'_{i_p}(x); \lambda)$ $(i=1,2; p=1,2,\ldots,P)$ in such a way that (0.11), (0.12), and (0.19) hold. The functions $\sigma_i(x)$ and $x'_i(x)$ $(i=1,2,\ldots,K)$ exist and are smoothly varying in the tubular region $\overline{T}_i$ of $\overline{V}$ covered by the rays in $\mathfrak{F}_i$.

The rays in $\mathfrak{F}_i$ are an $(m-1)$-parameter family of straight lines represented by the equations

$$x = X_i(\sigma,\tau) = \nu(\tau)\sigma + X_i^0(\tau) \qquad (\sigma \geq 0, \tau \in D_i), \quad (0.26)$$

where $x = X_i^0(\tau)$ $(\tau \in D_i)$ is a parametric representation of S_i (D_i is a compact connected subset of the rectangular domain $\{\tau : \tau_1 \leq \tau < \tau_2\}$. The mapping (0.26) defines a coordinate transformation from the half-cylinder (half-strip) $\mathcal{S}_i = \{(\sigma,\tau) : \sigma \geq 0, \tau \in D_i\}$ onto a tubular zone $\overline{T}_i$ in R^3 (R^2). This transformation has a unique inverse $\sigma = \sigma_i(x)$, $\tau = \tau_i(x)$. The function $x'_i(x)$ is equal to $X_i^0(\tau_i(x))$, and $\sigma_i(x)$ satisfies the eikonal equation $|\nabla\sigma_i(x)|^2 = 1$. The rays in $\mathfrak{F}_i$ have the following properties:

$$X_i(\sigma,\tau) \in C^k(\mathcal{S}_i) \quad \text{if} \quad X_i^0(\tau) \in C^{k+1}(D_i), \quad (0.27)$$

(0.28) The Jacobian $J_i(\sigma,\tau) = \partial(X_i)/\partial(\sigma,\tau)$ is (strictly) positive on $\mathcal{S}_i$,

(0.29) $$\partial_\sigma X_i(\sigma,\tau)\cdot\partial_{\tau^j}X_i(\sigma,\tau)\equiv 0 \qquad (j=1,m-1;(\sigma,\tau)\in \mathcal{S}_i),$$

and

(0.30) $$|\partial_\sigma X_i(\sigma_i(x),\tau_i(x))|^2 = 1 \qquad (x\in S_i\cup\bar{T}_i)\ .$$

In the case that ∂V itself is convex $(K=1)$ the above relationships are true on each of a finite number of tubular subregions $\bar{T}_i\cup S_i$ whose union is $\bar{V}$, where the curves $\{x|x=X_i(0,\tau),\ \tau^j=\text{const.},\ \tau\in D_i\}$, $j=1,m-1$, are curves of constant principle curvature on S_i (Note $\cup S_i=\partial V$). The function $\sigma(x)=\sigma_i(x)$ $(x\in\bar{T}_i\cup S_i)$ is smooth and satisfies $|\nabla\sigma|^2=n(x)$ on $\bar{V}$ (see Chap. 2, Section 7).

If $n(x)\not\equiv 1$ the reciprocal relationship between the ray systems and the "index of refraction" is more complicated. As discussed in Chapter 2 the rays of the field $\mathfrak{F}_i$ corresponding to a patch S_i that is locally convex relative to $n^{\frac{1}{2}}(x)$ can be characterized as an $(m-1)$-parameter family of curves

(0.31) $$x = X_i(\sigma,\tau)\in C^2(\mathcal{S}_i)\ ,$$

such that $\mathcal{S}_i$ is given parametrically by

(0.32) $$x = X_i(0,\tau) \qquad (\tau\in D_i)\ ,$$

and

(0.33) $$X_i(\sigma,\tau)\sim\nu(\tau)\sigma + X_i^0(\tau)$$ $$\partial_\sigma X_i(\sigma,\tau)\sim\nu(\tau)$$

as $\sigma\to\infty$, $\tau\in D_i$.

The transformation $x=X_i(\sigma,\tau)$ has the unique inverse $\sigma=\sigma_i(x)$, $\tau=\tau_i(x)$ for all $x\in\bar{T}_i=X_i(\mathcal{S}_i)$.

Furthermore, if $x\in\bar{T}_i$

(0.34) $$x_i'(x) = X_i(0,\tau_i(x))\ ,$$

(0.35) $$\partial_\sigma X_i(\sigma_i(x),\tau_i(x))\cdot\partial_{\tau^j}X_i(\sigma_i(x),\tau_i(x)) = 0\ ,$$

(0.36) $$n^{\frac{1}{2}}(x) = [|\partial_\sigma X_i(\sigma_i(x),\tau_i(x))|]^{-1}\ ,$$

(0.37) $$|\nabla\sigma_i(x)|^2 = n(x)\ ,$$

(0.38) $$\partial_\sigma X_i(\sigma_i(x),\tau_i(x)) = \nabla\sigma_i(x)/n(x)\ .$$

One can take the viewpoint that K normal congruences $\mathfrak{F}_1,\ldots,\mathfrak{F}_K$ of rays defining simple coverings of the infinitely long tubes $\bar{T}_i$ are given and that $n^{\frac{1}{2}}(x)$ is defined by (0.36) or (0.37) on $\bar{T}_i$ $(i=1,\ldots,K)$. The asymptotic behavior as $r\to\infty$ and the smoothness of $n(x)$ on $\bar{T}_i$ is then determined by the smoothness and asymptotic behavior as $\sigma \to \infty$ of $\partial_\sigma X_i(\sigma,\tau)$ as described in Chapter 2. Then we must require that the index defined by $\mathfrak{F}_i$ on T_i agrees with the index defined by $\mathfrak{F}_j$ $(j \neq i)$ on $\bar{T}_i \cap \bar{T}_j$ if $\bar{T}_i \cap \bar{T}_j \neq \emptyset$. On any portion of $\bar{V}$ not covered by any of the ray fields $\mathfrak{F}_i$ we are free to define $n(x)$ arbitrarily provided the index induced on $\bar{V}$ tends to one as $r \to \infty$ and is smooth; see Fig. 0.5.

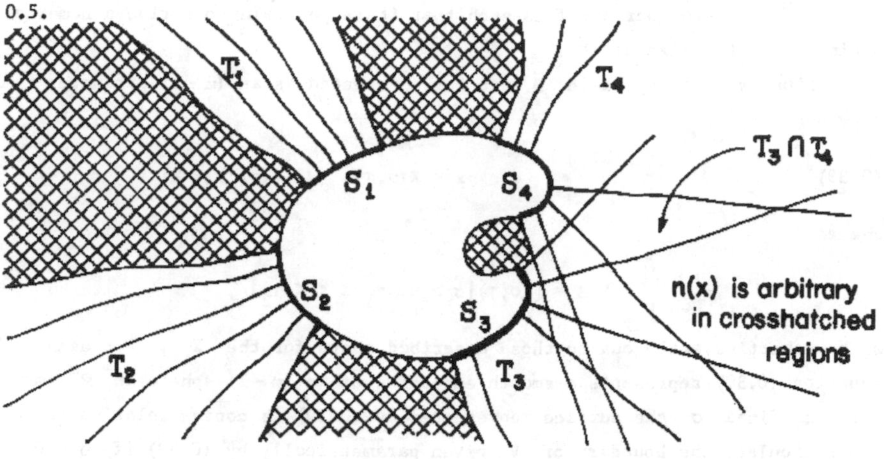

Figure 0.5

On the other hand, if one assumes that $n(x)$ is given, a more difficult problem arises: to establish the existence of a general class of surfaces S_i (curves) in $\mathbb{R}^3$ $(\mathbb{R}^2)$ that are locally convex or convex relative to $n^{\frac{1}{2}}(x)$. We address ourselves to this problem in Chapter 2.

By definition, S_i is locally convex relative to $n^{\frac{1}{2}}(x)$ if and only if it is an orthogonal surface (curve) in $\mathbb{R}^3$ $(\mathbb{R}^2)$ of the ray field $\mathfrak{F}_i$ that covers $\bar{T}_i \cup S_i$. If $n^{\frac{1}{2}}(x) \to 1$ as $r \to \infty$ (as we have assumed) it seems reasonable to expect that such a field $\mathfrak{F}_i$ exists on an infinitely long tubular region contained in some neighborhood of infinity, and that each ray (geodesic) of $\mathfrak{F}_i$ is asymptotic to a straight line as $r \to \infty$. Similarly, if $n^{\frac{1}{2}}(x)$ as $r \to \infty$, there should exist full neighborhoods of infinity covered by normally congruent fields of rays, whose orthogonal surfaces (curves) are, by definition, convex relative to $n^{\frac{1}{2}}(x)$. In Chapter 2 we prove, under physically reasonable hypotheses on the smoothness of n and the rate at which $n^{\frac{1}{2}}(x) - 1$ approaches zero as $r \to \infty$, that corresponding to each normal congruence $\hat{\mathfrak{F}}_i$ of straight line rays that defines a simple covering of some unbounded sectorial region $\hat{T}_i$ in $\mathbb{R}^m$ $(m = 2,3)$, there exists a unique ray field $\mathfrak{F}_i$ defined on an infinitely long tubular region $\bar{T}_i$ in $\mathbb{R}^m$ such that each ray in $\mathfrak{F}_i$ is asymptotic to a straight line ray in $\hat{\mathfrak{F}}_i$ as $r \to \infty$. The rays in $\mathfrak{F}_i$ are orthotomic relative to S_i; see [18, p. 105]. If careful track is kept of various constants that arise in our proof, specific numerical estimates of the sizes of the tubular regions covered by the rays in the corresponding $\mathfrak{F}_i$'s as functions of the magnitude of $n^{\frac{1}{2}}(x) - 1$ and its derivatives can be obtained.

Furthermore, using the above results, we prove in Chapter 2 that, corresponding to each ray field $\hat{\mathfrak{F}}$ of straight line rays emanating normally from a closed convex surface (curve) in $\mathbb{R}^3$ ($\mathbb{R}^2$), there exists a field $\tilde{\mathfrak{F}}$ of optical paths covering a neighborhood $\mathfrak{n}$ of infinity such that each ray in $\tilde{\mathfrak{F}}$ is asymptotic to a ray in $\hat{\mathfrak{F}}$. Again, our proof is such that it is possible to derive a numerical estimate for the size of $\mathfrak{n}$.

Finally, the ray field $\tilde{\mathfrak{F}}$ can be represented as an $(m-1)$ parameter family of curves

$$x = X(\sigma,\tau) \qquad ((\sigma,\tau) \in \mathcal{S}) , \tag{0.39}$$

where

$$\mathcal{S} = \{(\sigma,\tau) \mid \sigma \geq 0 , \tau_1 \leq \tau < \tau_2\} ,$$

with properties analogous to those described above for the X_i. For each $\sigma \geq 0$ equation (0.39) represents a smooth embedding of an $(m-1)$-sphere in $\mathbb{R}^m$ $(m = 2,3)$. For each fixed σ the surface represented by (0.39) is convex relative to $n^{\frac{1}{2}}(x)$. In particular, the boundary of V given parametrically by (0.39) if $\sigma = 0$ is convex relative to $n^{\frac{1}{2}}(x)$. Of course, in $\mathbb{R}^3$, while a smooth eikonal function $\sigma(x)$ is defined on $\mathfrak{n}$, there exists no globally defined smooth $\tau(x)$ on $\mathfrak{n}$ such that

$$x = X(\sigma(x) , \tau(x)) \qquad (x \in \mathfrak{n}) .$$

In Chapter 3 we present an algorithm for constructing $u_M(x,\lambda)$ in the case that ∂V is convex relative to $n^{\frac{1}{2}}(x)$. We set

$$u_M(x,\lambda) = e^{i\lambda\chi(x)}\Sigma_{j=0}^{M+1} \frac{A^j(x)}{\lambda^j} + v_M(x,\lambda) , \tag{0.40}$$

where

$$v_M(x,\lambda) = \Sigma_{j=2}^{M'} B^{j-2}(x)\lambda^{-j} . \tag{0.41}$$

$$(M' = M+2 \text{ if } M \text{ is odd} ; M' = M \text{ if } M \text{ is even}) ,$$

and then apply the differential operator L to u_M. If (i) we set $\chi(x) = \sigma(x) \equiv \sigma_1(x)$ $(x \in \overline{V} \equiv \overline{\mathfrak{n}})$, then $\chi(x)$ is a smooth solution of the eikonal equation

$$|\nabla\chi|^2 = n(x) \qquad (x \in \overline{V}) . \tag{0.42}$$

We suppose that (ii) there exist functions $A^0(x),\ldots,A^{M+1}(x)$ in $C^2(\overline{V})$ such that

$$(0.43)\qquad 2\nabla\sigma\cdot\nabla A^{0}(x)+\Delta\sigma(x)A^{0}(x)\equiv 0\,,$$

$$(0.44)\qquad 2\nabla\sigma(x)\cdot\nabla A^{j}(x)+\Delta\sigma(x)A^{j}(x)\equiv i\Delta A^{j-1}(x)\quad (j=1,\dots,M+1)$$

for all $x\in\bar{V}$, and (iii) there exist functions $B^{j-2}(x)$ ($j=0,\dots,M'$, where $M'=M+2$ if M is odd and $M'=M$ if M is even) in $C^{2}(\bar{V})$ such that

$$(0.45)\qquad B^{0}(x)\equiv\frac{f_{0}(x)}{n(x)}\,,\quad B^{1}(x)\equiv 0\,,\quad B^{j}(x)=-\frac{\Delta B^{j-2}(x)}{n(x)}$$

$$(j=2,\dots,M'-2\,;\ x\in\bar{V})$$

and with B^{M-1} defined to be zero if M is even. Assuming (i), (ii), and (iii), one concludes that

$$(0.46)\qquad Lu_{M}=f_{0}(x)+R(x,\lambda)\,,$$

where

$$R(x,\lambda)=e^{i\lambda\sigma(x)}\Delta A^{M+1}(x)\lambda^{-M-1}+\Delta B^{M'-3}(x)\lambda^{1-M'}+\Delta B^{M'-2}\lambda^{-M'}$$

$$(M'=M+2\ \text{if}\ M\ \text{is odd};\ M'=M\ \text{if}\ M\ \text{is even})\,.$$

Note that $R(x,\lambda)=\mathcal{O}(\lambda^{-M})$ whether M is even or odd. Condition (0.19) will be satisfied if we can show that ΔA^{j}, ΔB^{j} lie in $C(\bar{V})$ and

$$\Delta A^{j}\,,\ \Delta B^{j}=\mathcal{O}(r^{-(m+3)/2})\quad\text{as}\quad r\to\infty$$

uniformly on $\bar{V}$. Since by definition $\sigma(x)=0$ on ∂V, the boundary condition (0.11) is satisfied if

$$(0.47)\qquad A^{0}(x)=u_{0}(x)\qquad (x\in\partial V)$$

and

$$(0.48)\qquad A^{j}(x)+B^{j-2}(x)=0\qquad (j=0,\dots,M+1\,;\ x\in\partial V)\,.$$

The radiation condition (0.12) will be satisfied if

$$(0.49)\qquad \lim_{R\to\infty}\int_{r=R} r|h|^{2}dS=0\,,$$

for h equal to any of the following functions:

$$h=A_{r}+i\lambda(\sigma_{r}-1)A+\frac{(m-1)}{2r}A\,,$$

where $$A = \Sigma_1^{M+1} \lambda^{-j} A^j(x) ,$$

$$h = B^0 , B_r^{M'-2} + \frac{(m-1)}{2r} B^{M'-2} ,$$

$$h = B_r^{j-1} + \frac{(m-1)}{2r} B^{j-1} - iB^j \qquad (j = 1,\ldots,M'-2) ,$$

where $M' = M + 2 , M$ odd; $M' = M , M$ even.

The differential equation (0.43) and the initial condition (0.47) are satisfied by the function

$$A^0(x) = \frac{u_0(x'(x))}{\mu(\sigma(x) , x'(x))} , \tag{0.50}$$

where

$$\mu(\sigma(x) , x'(x)) = \exp\left\{\frac{1}{2}\int_0^{\sigma(x)} \frac{\Delta\sigma(X(\sigma , \tau(x'(x))))}{n(X(\sigma , \tau(x'(x))))} d\sigma\right\} ,$$

and

$$x = X(\sigma, \tau) \qquad (\sigma, \tau) \in \mathcal{S}$$

is the above mentioned representation of the rays in $\tilde{\mathfrak{J}}$. It is shown in [12] that

$$\mu(\sigma(x) , x'(x)) = C(x , x'(x)) = [K(x'(x))/K(x)]^{\frac{1}{2}} ;$$

see page 8 of this Introduction.

To verify that $A^0(x)$ as given by (0.50) satisfies (0.43) we first make use of the relationship

$$X_\sigma(\sigma(x) , \tau(x)) = \nabla\sigma(x)/n(x) , \tag{0.51}$$

which holds for all $x \in \overline{V}$, and the identity

$$x = X(\sigma(x) , \tau(x)) = X(\sigma(x) , \tau(x'(x))) ,$$

which holds for all $x \in \overline{V}$, to rewrite (0.43) as

$$\left[X_\sigma(\sigma , \tau) \cdot \nabla A^0(X(\sigma , \tau)) + \frac{\Delta\sigma(X(\sigma , \tau))}{2n(X(\sigma , \tau))} A^0(X(\sigma , \tau))\right]\Bigg|_{(\sigma , \tau)=(\sigma(x) , \tau(x'(x)))}$$

$$= \frac{d}{d\sigma} A^0(X(\sigma, \tau(x'(x))))\mu(\sigma , x'(x))\Bigg|_{\sigma=\sigma(x)} = 0 .$$

Therefore, if we set

$$(0.52) \qquad A^0(X(\sigma, \tau(x'(x)))) = \frac{u_0(x'(x))}{\mu(\sigma, x'(x))},$$

equation (0.43) and the initial condition (0.47) are satisfied if $\sigma = \sigma(x)$.

Similarly, it can easily be verified that the transport equations (0.44) and the initial conditions (0.48) are equivalent to the integral equations

$$(0.53) \qquad A^{j+1}(x) = -B^{j-1}(x'(x))/\mu(\sigma(x), x'(x))$$

$$+ \frac{i}{2\mu(\sigma(x), x'(x))} \int_0^{\sigma(x)} \Delta A^j (X(\sigma, \tau(x'(x))))\mu(\sigma, x'(x))d\sigma$$

$$(j = 0, \ldots, M),$$

where $B^{-1} \equiv 0$. Note that $A^j(x) = A^j(X(\sigma(x), \tau(x'(x))))$, so that the $A^j(x)$ can be constructed by solving

$$(0.54) \qquad A^{j+1}(X(\sigma, \tau)) = -B^{j-1}(X(0, \tau))/\mu(\sigma, X(0, \tau))$$

$$+ \frac{i}{2\mu(\sigma, X(0, \tau))} \int_0^{\sigma} \mu(s, X(0, \tau)) \Delta A^j(X(s, \tau))ds$$

recursively for $A^1(X(\sigma, \tau)), \ldots, A^{M+1}(X(\sigma, \tau))$ and then setting $(\sigma, \tau) = (\sigma(x), \tau(x'(x)))$ recalling that $x'(x) = X(0, \tau(x))$.

In Chapter 3 we show by an induction argument that the $B^j(x)$ $(j = 0, \ldots, M' - 2)$ as determined recursively by (0.45), and the $A^j(x)$ $(j = 1, \ldots, M+1)$ as determined recursively by (0.54) are continuous together with their 2nd derivatives on $\overline{V}$, satisfy the radiation condition (0.49) and that

$$(0.55) \qquad \Delta B^j(x), \ \Delta A^j(x) = \mathcal{O}(r^{-(m+3)/2}) \qquad (r \to \infty, m = 2 \text{ or } 3),$$

as is required to satisfy (0.19). We obtain these results under the following hypotheses on $n(x)$:

$$(0.56) \qquad n \in C^{2M+6}(\mathbb{R}^m) \qquad (m = 2,3),$$

$$0.57) \qquad n(x) \geq n_0,$$

$$(0.58) \qquad \sup_{\mathbb{R}^m} r^2 |n(x) - 1| < \infty,$$

$$(0.59) \qquad \sup_{\mathbb{R}^m} r^{2+|p|} |D^p n(x)| < \infty \ (1 \leq |p| \leq 2M + 6),$$

where $p = (p_1, \ldots, p_m)$, $|p| = \Sigma_1^m p_i$ and the p_i are non-negative integers.

We also hypothesize that

$$g(x) \in C^{2M+4}(\partial V), \tag{0.60}$$

$$f(x) \in C^{2M}(\overline{V}), \tag{0.61}$$

$$D^p f(x) = \mathcal{O}(r^{-|p|}) \quad (|p| \le M, \ x \in \overline{V}). \tag{0.62}$$

The approximate solution (0.40) is consistent in form with (0.21) - (0.24) if we set $p = 1$ in (0.21) and (0.23) and assume that

$$A^{i1}(\sigma, x'(x); \lambda) \cong \Sigma_{j=0}^{M+1} \frac{A^{ij1}(\sigma, x'(x))}{\lambda^j} \quad (i = 1,2), \tag{0.63}$$

and

$$B(x, \lambda) \cong v_M(x, \lambda). \tag{0.64}$$

Essentially the same argument used in Chapter 3 to construct the approximate solution $u_M(x, \lambda)$ can be applied to the construction of $u_M(x, \lambda)$ in the case that the support of $g(x)$ and $f_0(x)$ intersects a finite number of disjoint patches $S_1, S_2, \ldots, S_K$ on ∂V that are locally convex relative to $n^{\frac{1}{2}}(x)$ if $\overline{T}_i \cap S_j \equiv 0$ and $\overline{T}_i \cap (\partial V - \cup_1^K S_i) = 0$ $(i,j = 1,2,\ldots,K)$. (See Fig. (0.6)). If an

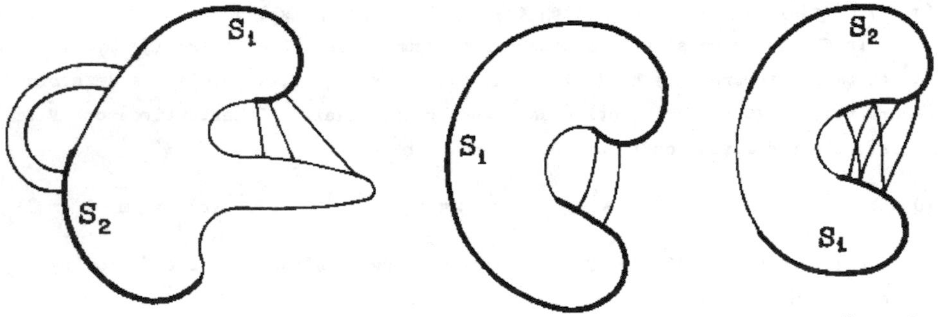

Figure 0.6

"Ansatz" of the form

$$u_M(x, \lambda) = \Sigma_{k=0}^{K} e^{i\lambda\sigma_k(x)} \Sigma_{j=0}^{M+1} \frac{A^{jk}(x)}{\lambda^j} + v_M(x, \lambda), \tag{0.65}$$

$$v_M(x, \lambda) \equiv \Sigma_{j=1}^{M'} \frac{B^{j-2}(x)}{\lambda^j}$$

is made, then (0.19), the radiation condition (0.12) and the boundary condition (0.11) will be satisfied on $\overline{V}$ if (i) the $B^{j-2}(x)$ satisfy (0.45) on $\overline{V}$,

(ii) for $x \in \overline{T}_k$

$$A^{0k}(x) = u_0(x_k'(x))/\mu(\sigma_k(x), x_k'(x)),$$

(iii) the A^{jk} $(j = 1,\ldots M+1)$ satisfy the integral equations (0.53) and (0.54) on $\overline{T}_k$ with $x'(x)$ replaced by $x_k'(x)$ and with $\sigma = \sigma_k(x)$ (note that $A^{jk}(x) = 0$ on $\partial T_k - S_k$, and that $A^{jk}(x) \in C_0^2(\overline{T}_k)$), and (iv) $A^{jk}(x) \equiv 0$ if $x \in \overline{V} - (T_k \cup S_k)$.

We remark that (0.65) is consistent with (0.21) - (0.24) if we set

$$A^{ik}(\sigma_k(x), x_k'(x); \lambda) \cong \Sigma_{j=1}^{M+1} A^{ijk}(\sigma_k(x), x_k'(x))\lambda^{-j}, \tag{0.66}$$

$$B(x, \lambda) \cong v_M(x,\lambda) \tag{0.67}$$

$$(i = 1, 2; k = 0,\ldots,K; x \in \overline{V})$$

and if we notice that $u_M(x,\lambda)$ in (0.65) is equal to

$$\Sigma_{p=1}^{P} e^{i\lambda\sigma_{i_p}(x)} \Sigma_{j=0}^{M+1} A^{ji_p}(x)\lambda^{-j} + v_M(x, \lambda) \tag{0.68}$$

$$(1 \le P \le K)$$

if $x \in \overline{T}_i$ for at least one value of i and that $u_M(x, \lambda)$ is identically zero if $x \in \overline{V} - \overline{V}_K$, where $\overline{V}_k = \cup_1^K \overline{T}_i$. (Recall the definition of the index i_p given in the discussion after (0.21).)

Similar considerations apply to the construction of an approximate solution u_M^* of (0.10), (0.12), and (0.17).

The optical paths on the ray field $\mathfrak{F}_1 \equiv \tilde{\mathfrak{F}}$ that corresponds to a closed surface ∂V that is convex relative to $n^{\frac{1}{2}}$ are free of caustics; see [8, p. 35]. In this case examination of the terms of the approximate (geometrical optics) solution $U_M^1(x, \lambda)$ given by (0.40) if $f_0(x) \equiv 0$ and the approximate (geometrical optics) solution $U_M^2(x, \lambda)$ given by (0.40) if $u_0(x) \equiv 0$ leads to the following conclusion. The function U_M^1 is determined by the values of $x'^j(x)$ $(j = 1,\ldots,m)$, where the x'^j are the coordinates of x', $u_0(x'(x))$, and derivatives of u_0 at $x'(x)$. Also U_M^2 is determined by the values of $x'^j(x)$ $(j = 1,\ldots,m)$, $f_0(x'(x))$, $f_0(x)$, $n(x)$ and the derivatives of these functions. (Recall that $x'(x)$ is the point on ∂V closest to x in the sense of the Riemannian metric $ds = n^{\frac{1}{2}}(x)|dx|$; alternatively, $x'(x)$ is the point of ∂V on the optical path of least time from x to ∂V.) Similar conclusions can be drawn about U_M^{*1} and U_M^{*2}.

If ∂V can be illuminated from the exterior, then (0.18) holds; and we draw the further conclusions that to within an error that is algebraically small with respect to λ, the scalar field $U^1(x, \lambda)$ at a typical point x of $\overline{V}$ is determined by the values of $x'^j(x)$ $(j = 1,\ldots,m)$, $u_0(x'(x))$, and derivatives of these functions, while $U^2(x, \lambda)$ at a typical point x of $\overline{V}$ is determined by the values of $x'^j(x)$ $(j=1,2,\ldots,m)$, $f_0(x'(x))$, $f_0(x)$, $n(x)$ and derivatives of these functions. (Note that the approximation we get for $U^2(x, \lambda)$ in the case $f(x') \equiv 0$ $(x' \in \partial V)$ holds even if ∂V is not convex relative to $n^{\frac{1}{2}}(x)$) .

In cases where ∂V consists of a finite number of disjoint patches $S_i(i = 1,2,\ldots,K)$ (containing the support of $g(x')$ and $f(x')$, $x' \in \partial V$) that are locally convex relative to $n^{\frac{1}{2}}(x)$, joined together by smooth seams, the rays of the field $\mathfrak{F}_i$ that corresponds to S_i may intersect the rays of one or more of the ray fields $\mathfrak{F}_j(j \neq i)$ that correspond to S_j ; see Fig. 0.6. In this case weak focussing of the energy of the geometrical optics field does occur. If the tube $\overline{T}_i$ covered by the rays in $\mathfrak{F}_i$ does not intersect $S_j(j \neq i)$ or $\partial V - \cup_{i=1}^K S_i$, then the approximate field at x, is given by (0.65) and is a superposition of P waves propogated from the points $x_i^!(x)$ on S_i $(p = 1,2.,,,.P)$, where the optical paths of least time from x terminate. Examination of the terms of the approximate solution $U_M^1(x, \lambda)$ given by (0.65) if $f_0(x) \equiv 0$, leads to the conclusion that $U_M^1(x, \lambda)$ is determined by the values of $x_{i_p}^{!j}(x)$ $(j = 1,2,\ldots,m)$, $u_0(x_{i_p}^!(x))$ $(p = 1,2,\ldots,P)$, and derivatives of these functions. Examination of the terms of the approximate solution $U_M^2(x, \lambda)$ given by (0.65) if $u_0(x) \equiv 0$ on ∂V leads to the conclusion that the approximate solution $U_M^2(x,\lambda)$ is determined by the values of $x_{i_p}^{!j}(x)$ $(j = 1,2,\ldots,m)$, $f_0(x_{i_p}^!(x))$ $(p = 1,2,\ldots,P)$, $f_0(x)$, $n(x)$ and derivatives of these functions. (Similar conclusions can be drawn about $U_M^{*j}(x, \lambda)$ $(j = 1,2)$.) If ∂V can be illuminated from the exterior, then (0.18) holds; and we draw the further conclusion that to within an algebraically small error the same conclusions can be drawn about the exact solutions $U^j(x, \lambda)$ $(j = 1,2)$.

If any of the rays in $\mathfrak{F}_i$ $(i = 1,2,\ldots,K)$ do intersect S_j for some $j \neq i$ (but not $\partial V - \cup_1^K S_k$), then (0.65) is no longer valid. But it should still be possible to construct an approximate solution of similar form to (0.65) provided that none of the rays from $\mathfrak{F}_i$ are tangentially incident on S_j . This would entail a proof that the subfield $\mathfrak{F}_{ij}$ of $\mathfrak{F}_i$ that obliquely intersects S_j is reflected into another subfield $\hat{\mathfrak{F}}_{ij}$. A subfield $\hat{\mathfrak{F}}_{ij\,j'}$ of $\hat{\mathfrak{F}}_{ij}$ may obliquely intersect $S_{j'}$ $(1 \leq j' \leq K)$. Again it would be necessary to prove that $\mathfrak{F}_{ij\,j'}$ is reflected into another subfield $\hat{\hat{\mathfrak{F}}}_{ij\,j'}$, etc. It would be necessary to require that after a finite number of reflections the remaining field of rays extends to infinity without intersecting ∂V . (See Fig. (0.7)). The argument we present in Chapter 2 can probably be modified to give the necessary proof.

We remark that the a priori estimates derived in Chapter 1, in particular the pointwise estimate (0.15), hold under conditions where the support of $f_0(x')$ or $g(x')$ $(x' \in \partial V)$ is neither locally convex relative to $n^{\frac{1}{2}}(x)$, nor consists of a finite number of disjoint locally convex patches. In such a situation the rays emanating normally from some connected subset of the support of $g(x')$ of $f_0(x')$ may be weakly or strongly focussed, or may be tangent to a caustic curve or surface

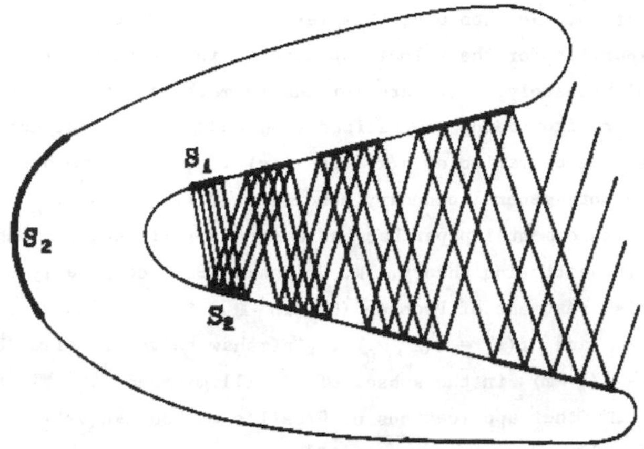

Figure 0.7

in $\tilde{V}$. Or it can happen that rays from the support of $g(x')$ or $f_0(x')$ intersect $\partial V - \bigcup_1^K S_i$ are tangent to ∂V. The algorithm presented in Chapter III for constructing $U_M^1(x, \lambda)$ or $U_M^2(x, \lambda)$ fails in such cases. However, our estimate (0.15) for $|u(x, \lambda)|$ provides an upper bound on the strength of the field $U^1(x, \lambda)$ or $U^2(x, \lambda)$ for $\lambda \gg 1$. In particular, we are able to predict that the strength of the field $U^j(x, \lambda)$ $(j = 1,2)$ due to the sources distributed on the boundary is at most of the order of magnitude of $r^{(1-m)/2}\lambda^{(3+m)/2}$ as $\lambda \to \infty$ at every point of $\bar{V}$. The strength of the field $U^2(x, \lambda)$ due to the source distribution $f(x)$ is at most of the order of magnitude of $r^{(1-m)/2}\lambda^{(1+m)/2}$.

To obtain more precise information about the behavior of the field $u(x, \lambda)$ in the above mentioned cases it would be necessary to construct a sufficiently accurate approximate solution $u_A(x, \lambda)$ of problem (P), and then to apply the pointwise estimate (0.15) to $u(x, \lambda) - u_A(x, \lambda)$.

In a limited number of special cases, for example, if $n(x)$ is radially symmetric and ∂V is a sphere or if ∂V is a circular cylinder and $n(x)$ is radially symmetric, it is possible to obtain an asymptotic expansion for the exact solution u from a series or integral representation. In all but special cases an asymptotic approximation has to be obtained directly, the guide for its form being that of an appropriate special case. For example, see [7, 10, 11].

F. Ursell [19] has solved Problem (U) in $\mathbb{R}^2$ for the case $n(x) \equiv 1$, $\alpha \equiv 0$, $\beta \equiv 1$, $f(x) \equiv 0$, and ∂V convex. In Ursell's work $u(x, \lambda)$ is interpreted as the amplitude of the time-harmonic component of the velocity potential of the acoustic field produced by a double layer distribution of radiating sources on the boundary of a slightly compressible homogeneous medium. He cleverly constructed a Fredholm integral equation for the boundary values of $u(x, \lambda)$ with a kernel that tends to zero as $\lambda \to \infty$. He proved that the leading term of the Neumann series for this integral equation is an asymptotic approximation for the boundary values of the exact solution. He then used this result to derive the leading term of an asymptotic expansion for the velocity potential in the exterior of the radiating obstacle. Unfortunately, there are formidable mathematical obstacles that prevent his approach form from being generalized to higher dimensions, nonconvex boundaries ∂V or to the case of variable $n(x)$ or $E(x)$. Ursell's method can probably be adapted to the more general boundary condition (0.17) in $\mathbb{R}^2$.

Ursell's fundamental paper has had a strong influence on much of the subsequent work in scattering theory. R. Grimshaw [9] successfully used Ursell's approach in his treatment of Problem (U) in $\mathbb{R}^2$ for ∂V convex, $n(x) \equiv 1$, $\alpha \equiv 0$, $\beta \equiv 1$, $g(x) \equiv 0$, and $f(x) = \delta(x, x_0)$. Grimshaw solved Problem (U) for integer values of $M - \frac{1}{2}(1+m)$ in the subset of $\overline{V}$ illuminated directly by the point source at x_0. Other applications of Ursell's method can be found in the work of V. M. Babič [1, 2] and L. A. Muraveĭ [17].

It remains an open problem to achieve one of Ursell's original goals: to solve Problem (U) in $\mathbb{R}^3$ if $\beta(x) \not\equiv 0$ in the boundary condition (0.17), at least in the case $\alpha \equiv 0$, $\beta(x) \equiv 1$, $n(x) \equiv 1$, $f_0(x) \equiv 0$ and ∂V convex.

It is also an open problem to separate and describe the contributions to the total field of diffracted radiation or radiation from subsets of $\overline{V}$ where derivatives of g, f, or n are discontinuous. It is because of the absence of a clearly defined shadow boundary that the asymptotic solution $u_M(x, \lambda)$ of Problem (U) which we construct has a relatively simple form. Diffraction effects and secondary radiation effects are absorbed into the error terms.

C. S. Morawetz and D. Ludwig [16] have considered Problem (U) for $m = 2,3$ in the case $n(x) \equiv 1$, ∂V convex, $\alpha \equiv 1$, $\beta \equiv 0$, $g \equiv 0$, and $f(x) = \delta(x, x_0)$. Their approach is to first construct a function u_A that (i) vanishes on ∂V, (ii) satisfies the radiation condition (0.12), and (iii) is an approximate solution of the reduced wave equation on $\overline{V}$. They derive a priori bounds similar to (0.13) - (0.15) that hold for solutions of Problem (P) in the case that ∂V is star-shaped and $n(x) \equiv 1$, and they apply them to obtain an estimate for the difference between the exact solution u and u_A. They thus prove that u_A is an asymptotic approximation of u in the illuminated portion of $\overline{V}$.

References for Chapter 0

1. V. M. Babič, On the short-wave asymptotic behavior of the Green's function for the exterior of a bounded, convex region. Dokl. Akad. Nauk SSSR 146 (1962), 571-573.
2. V. M. Babič, The asymptotic behavior of the Green's function of certain wave problems. I. The stationary case. Mat. Sbornik (N.S.) 86 (128) (1971), 518-537.
3. C. O. Bloom, A rate of approach to the steady state of solutions of second-order hyperbolic equations. J. Differential Equations 19 (1975), 296-329.
4. R. N. Buchal, The approach to steady state of solutions of exterior boundary-value problems for the wave equations, J. of Math and Mech. 12 (1963), 225-234.

5. D. M. Èĭdus, On the principle of limiting absorption, Mat. Sbornik N.S. 57 (99) (1962) 11-34. = A.M.S. Transl. (2) 47 (1965), 157-191.

6. D. M. Èĭdus, Some boundary-value problems in infinite regions, Izv. Akad. Nauk SSSR. Ser. Mat. 27 (1963), 1055-1080. = A.M.S. Transl. (2) 53 (1966), 139-168.

7. W. Franz, Ueber die Greenschen Funktionen des Zylinders und der Kugel. Z. Naturforsch. 9a (1954), 705-716.

8. F. G. Friedlander, Sound Pulses, Cambridge Univ. Press, London, 1958.

9. R. Grimshaw, High-frequency scattering by finite convex regions, Comm. Pure Appl. Math XIX (1966), 167-198.

10. D. S. Jones, High-frequency refraction and diffraction in general media, Philos. Trans. Roy. Soc. London Ser. A 255 (1962/63), 341-387.

11. J. B. Keller and B. R. Levy, Diffraction by a spheriod, Canad. J. Phys. 38 (1960), 128-144.

12. M. Kline and I. Kay, Electromagnetic Theory and Geometrical Optics, Interscience, New York, 1965.

13. P. Lax and R. Phillips, Scattering Theory, Academic Press, New York, 1967.

14. R. K. Luneberg, Mathematical Theory of Optics, Univ. Calif. Press, Berkeley, 1964.

15. C. S. Morawetz, The limiting amplitude principle, Comm. Pure and Appl. Math XV (1962), 349-361.

16. C. S. Morawetz and D. Ludwig, An inequality for the reduced wave operator and the justification of geometrical optics, Comm. Pure Appl. Math. 21 (1968), 187-203.

17. L. A. Muraveĭ, The decrease of solutions of the second exterior boundary-value problem for the wave equation with two space variables, Dokl. Akad. Nauk SSSR. 193 (1970), 996-999.

18. O. N. Stavroudis, The Optics of Rays Wavefronts and Caustics, Academic Press, New York, 1972.

19. F. Ursell, On the short-wave asymptotic theory of the wave equation $(\nabla^2 + k^2)\varphi = 0$, Proc. Camb. Philos. Soc. 53 (1957), 115-133.

CHAPTER 1

A PRIORI BOUNDS

1. Introduction

In this Chapter we obtain a priori estimates for the solution $u(x, \lambda)$ of the following radiation problem:

$$(1.1) \qquad Lu = f(x, \lambda) \quad (x \in V \subset \mathbb{R}^m,\ m = 2, 3; \lambda > 0)$$

$$(P) \qquad (1.2) \qquad u|_{\partial V} = u_0(x) \quad (x \in \partial V)$$

$$(1.3) \qquad \lim_{R\to\infty} \int_{r=R} r|\mathcal{D}_1 u|^2 = 0 \quad (r = (x \cdot x)^{\frac{1}{2}}) ,$$

where

$$Lu = \Delta u + \lambda^2 n(x) u ,$$

and

$$\mathcal{D}_1 u = u_r - i\lambda u + \frac{(m-1)}{2r} u \quad (m = 2,3) .$$

If $m = 2(3)$, then V is the exterior of a smooth closed curve (surface) ∂V. We assume that ∂V can be illuminated from the exterior by a convex curve (surface) contained in V (see Definition 2.1 of Section 2).

We require that

$$\text{(i)} \quad u_0(x) \in C^1(\partial V) , \qquad \text{(ii)} \quad n(x) \in C^2(\overline{V}) ,$$

$$(H) \qquad \text{(iii)} \quad f(x,\lambda) \in C(\overline{V}) \ \underline{\text{for every}}\ \lambda > 0 ,$$

$$\text{(iv)} \quad \int_V r^2 |f|^2 < \infty , \qquad \text{(v)} \quad n(x) \geq n_0 > 0 \ \underline{\text{for all}}\ x \in \overline{V} .$$

Here $\overline{V} = V \cup \partial V$. In addition we require that <u>as</u> $r \to \infty$

$$\text{(vi)} \quad \left|\frac{1}{n(x)} - 1\right| = \mathcal{O}(r^{-p}) \ \underline{\text{for some}}\ p > 2 ,$$

$$(H) \qquad \text{(vii)} \quad \nabla n(x) = \mathcal{O}(r^{-2}) ,$$

$$\text{(viii)} \quad \partial^{i+j} n(x) / \partial x^i \partial x^j = \mathcal{O}(r^{-3}) \quad (i + j = 2,\ i, j \geq 1) .$$

Most of this Chapter is devoted to obtaining estimates for the "energy norms"

$$\|u_{\nu^*}\|_{\partial V} = \left(\int_{\partial V} |\nu^* \cdot \nabla u|^2\right)^{\frac{1}{2}} ,$$

where ν^* in the exterior unit normal to ∂V,

$$\left\|\frac{\nabla u}{r}\right\|_V = \left(\int_V \frac{|\nabla u|^2}{r^2}\right)^{\frac{1}{2}} ,$$

and

$$\left\|\frac{u}{r}\right\|_V = \left(\int_V \frac{|u|^2}{r^2}\right)^{\frac{1}{2}} ,$$

in terms of the given boundary data u_0 and the source term f. We show how to compute the constants involved in these a priori estimates explicitly in terms of ∂V, and n. The estimates hold as $\lambda \to \infty$. We use these L_2-estimates to derive an a priori estimate for the scalar field strength $|u(x,\lambda)|$ in terms of f and u_0 that holds uniformly on $\overline{V}$ as $\lambda \to \infty$.

As me mentioned in Chapter 0, D. Ludwig and C. S. Morawetz [4] derived a priori bounds that hold for solutions of Problem P in the case that ∂V is star-shaped and $n(x) \equiv 1$, and they applied them to obtain an estimate for the difference between the exact solution and an approximate solution $u_A(x,\lambda)$ that establishes $u_A(x,\lambda)$ as an asymptotic approximation of $u(x,\lambda)$ in the "illuminated" portion of $\overline{V}$.

The a priori estimates obtained by Morawetz and Ludwig in [4] do not include an estimate for $(\|\nabla u/r\|_V)^2$. Such an estimate is obtained in a more recent paper of Morawetz [5] on energy decay for the wave equation, but it is obtained by an argument that is different from ours; see W. A. Strauss [6] for a related estimate.

Bloom [1] has derived a priori estimates for the solution of Problem P in the case of a general second-order elliptic operator, but only for star-shaped ∂V. Our pointwise estimate for $u(x,\lambda)$ improves upon Bloom's because it is uniform over $\overline{V}$, and because we allow a much wider class of scattering obstacles, e.g. "snake shaped" bodies. For real λ our local energy estimate generalizes a recent estimate of C. S. Morawetz [5] which holds for complex λ for an even wider class of obstacles in $\mathbb{R}^2$ and $\mathbb{R}^3$, but only for $n \equiv 1$. We also present a shorter, better motivated derivation of the basic divergence identity in [1].

Our derivation of the a priori estimates in Theorem 7.1 and Theorem 8.1 is similar in structure to Bloom's proof in [1]. There are significant differences between the choice of multipliers made here and in [1], and in the various delicate calculations and estimates. There are also several simplifications in our work resulting from the fact that we treat the differential operator $\Delta + \lambda^2 n(x)$ rather than a general self-adjoint operator with variable coefficients. In particular we are able to avoid the patching argument used in Appendix III of [1], and the complicated choice of $\rho(x)$ made in Appendix II of [1].

The multipliers we use are related to those defined in [2]. There we established decay rates for the local energy of solutions of the wave equation defined in exterior regions with boundaries that can be illuminated from the interior.

This Chapter is organized as follows. Section 2 deals with geometry. We describe there the coordinate system in terms of which the multipliers, and other auxiliary functions are defined.

In Section 3 we derive a differential inequality for functions $u \in C^1(\overline{V}) \cap C^2(V)$. This inequality is obtained from a basic divergence identity proved in [3]. Here we give an alternative derivation based upon a variational principle. We integrate this inequality over the region $V(R)$ bounded by ∂V, and the sphere $S(R) = \{x: |x| = R\}$. The divergence terms give rise to an integral over ∂V, and an integral over $S(R)$ of quadratic functions of u and first derivatives of u. The remaining terms obtained are integrals over $V(R)$. Our a priori estimates are derived from this inequality.

Sections 4 and 5 are devoted to carefully estimating the integrands of the integrals over $V(R)$. In Section 6 we prove that the integral over $S(R)$ is bounded from below by a function of R that tends to zero as $R \to \infty$, if u satisfies the Radiation Condition (1.3).

In Section 7 we apply these results to the integral inequality derived in Section 3. We assume that u is the solution of (P) and let $R \to \infty$. We obtain a preliminary estimate for $\|r^{-1}\nabla u\|_V$ and $\|\nu^* \cdot \nabla u\|_{\partial V}$ in terms of the boundary data u_0, the source term f, and $\|r^{-1}u\|_V$.

At this point it appears that little has been accomplished since $\|r^{-1}u\|_V$ is as yet an unknown quantity. We overcome this difficulty by establishing a "small multiples" estimate. This is an upper bound for $\|r^{-1}u\|_V$ in terms of norms of f, u_0 and small multiples of the unknown quantities $\|r^{-1}\nabla u\|_V$ and $\|\nu^* \cdot \nabla u\|_{\partial V}$, multiples that approach zero as $\lambda \to \infty$. We use this upper bound in the preliminary estimate to bound $\|\nu^* \cdot \nabla u\|_{\partial V}$ and $\|r^{-1}\nabla u\|_V$ from above in terms of norms of f and u_0. We then use this result in the small multiples inequality to obtain an upper bound for $\|r^{-1}u\|_V$ in terms of norms of f and u_0.

Finally, in Section 8 we use the estimates for $\|r^{-1}\nabla u\|_V$, $\|r^{-1}u\|_V$, and $\|\nu^* \cdot \nabla u\|_{\partial V}$ obtained in Section 7 in an integral representation for u to derive an a priori estimate for $|u(x, \lambda)|$ that holds uniformly on $\bar{V}$ as $\lambda \to \infty$.

We remark that the estimates of this Chapter imply a uniqueness theorem for the solution of Problem P if λ is sufficiently large. For if $f = u_0 = 0$, our estimate for $|u(x, \lambda)|$ reduces to $|u(x, \lambda)| \equiv 0$ for every $x \in \bar{V}$.

2. Geometric Preliminaries

Let C be a convex body in $\mathbb{R}^m$ ($m = 2$, or 3) with smooth boundary ∂C. Let ν and ν^* be the unit exterior normals to ∂C and to ∂V, respectively.

Definition 2.1. A scattering obstacle ∂V can be illuminated from the exterior if and only if there exist a $c_0 > 0$ and a convex body C, with $\partial V \subset C$, such that (i) if $X^0 \in \partial C$ and $x \in \partial V$ lie on the same interior normal to ∂C, then $\nu(X^0) \cdot \nu^*(x) \geq c_0$, and (ii) any two interior normals to ∂C intersect only after passing through ∂V.

Definition 2.1 means that each point of ∂V can be seen along one and only one interior normal to ∂C. An example of a scattering obstacle ∂V that can be illuminated from the exterior, but which is neither star-shaped nor illuminable from the interior is a "snake"; see Fig. 1. Henceforth ∂V will be an obstacle that can be illuminated from the exterior.

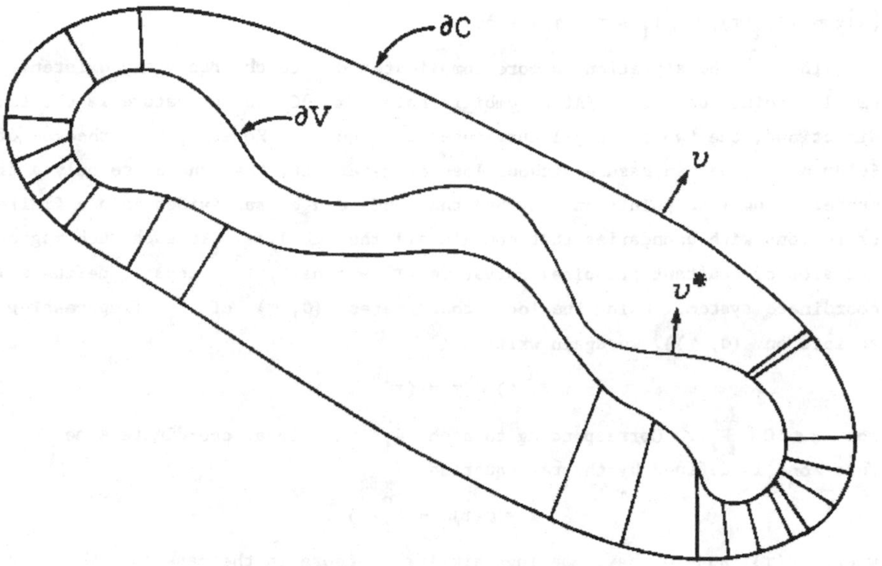

Figure 1

In $\mathbb{R}^2$ the normals to ∂C define a coordinate system on $\bar{V}$. Let τ be arc-length along the curve ∂C measured from some fixed point of ∂C, and let

$$x = X^0(\tau) \quad (\tau_1 \le \tau \le \tau_2)$$

be a representation of C, with $X^0(\tau_1) = X^0(\tau_2)$. The lines normal to ∂C are described by

(2.1) $$x = \nu(\tau)\sigma + X^0(\sigma) ,$$

where $\nu(\tau)$ is the unit exterior normal to ∂C at the point $X^0(\tau)$, and $|\sigma|$ measures distance from ∂C along this normal: $\sigma > 0$ in $\operatorname{ext}\partial C$, $\sigma = 0$ on ∂C, $\sigma < 0$ in $\operatorname{int}\partial C$. Equation (2.1) defines a coordinate system in $\bar{V}$. For each x in $\bar{V}$ there is a unique ordered pair $(\sigma(x), \tau(x))$ such that (2.1) holds. Further there exists a "half-strip" in (σ, τ)-space, call it $\mathcal{S}$, bounded by the curves $\tau = \tau_1$, $\tau = \tau_2$, $\sigma = \sigma_1(\tau)$ such

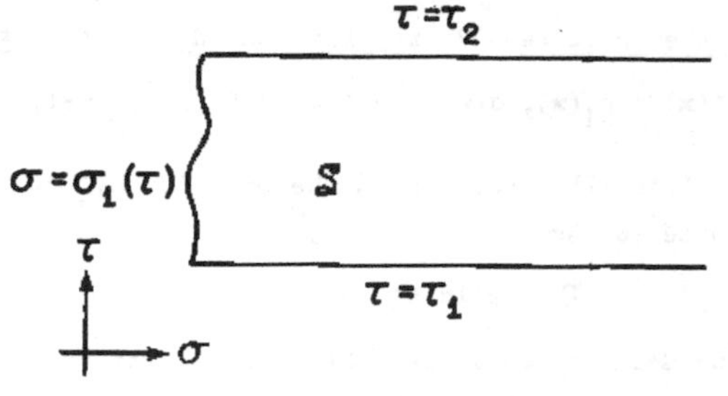

Figure 2

that (2.1) defines a smooth 1-1 mapping $x = X(\sigma, \tau)$ from $\mathcal{S}$ onto $\bar{V}$ and $\{x | x = x(\sigma_1(\tau), \tau), \tau_1 \le \tau \le \tau_2\} = \partial V$.

In $\mathbb{R}^3$ the situation is more complicated due to the necessary existence of umbilic points on ∂C . (At an umbilic point $x \in \partial C$ the curvature is the same in all directions; the two principal curvatures are equal.) However, by a theorem of E. A. Feldman [3], we can assume without loss of generality that there are only a finite number of umbilic points on ∂C and that ∂C can be subdivided into a finite number of regions with boundaries that contain all the umbilics. In each such region R_i the arcs of constant principal curvature τ^1 = const., τ^2 = const. define a local coordinate system. Using the local coordinates (σ, τ) of R_i (suppressing subscripts on (σ, τ)), we again write

$$x = X_i^0(\tau) \quad (\tau = (\tau^1, \tau^2))$$

for $x \in \partial C \cap R_i$. Corresponding to each R_i is a local coordinate zone V_i in $\bar{V}$. This zone is defined by the ray equation

$$x = \nu(\tau)\sigma + X_i^0(\tau) , \tag{2.2}$$

where $\nu(\tau)$ and σ have meanings similar to those in the case of $\mathbb{R}^2$. The zones V_i cover all of V except for the points that lie on normals emanating from the umbilics on ∂C . Again we associate with each x in a zone V_i the ray coordinates $(\sigma(x), \tau(x))$ (suppressing the subscript i on τ) that correspond to it through (2.2). Finally we denote the 1-1 mapping determined by (2.2), from a "half-cylinder" $\mathcal{S}_i$ in (σ,τ)-space onto $\bar{V}_i$ (minus the rays emanating from umbilics) by X_i; that is, for each $x \in \bar{V}_i$ (minus the rays emanating from umbilics)

$$x = X_i(\sigma(x)), \tau(x)) .$$

In view of its geometrical interpretation, the local coordinate function $\sigma(x)$ can be extended to a globally continuous function of x on all of V, while the local coordinate function τ cannot be so extended.

In $\mathbb{R}^2$ let $\rho_1(x)$ be the radius of curvature of ∂C at the point x' where the normal to ∂C passes through x . In $\mathbb{R}^3$ let $\rho_i(x)$ $(i = 1,2)$ be the principal radii of curvature of ∂C at the point x' where the normal to ∂C passes through x.

We make the following observations for use in the sequel.

__Lemma__ 2.1. *__If the obstacle ∂V is illuminated from the exterior by ∂C, then__*

$$\underset{\bar{V}}{\mathrm{Min}} \, [\sigma(x) + \rho_1(x), \sigma(x) + \rho_s(x)] > 0 \quad (s = 1, m-1; \; m = 2,3) .$$

__Proof.__ Since V is illuminated from the exterior by ∂C, every x in $\bar{V}$ lies on some coordinate surface

$$\Sigma = \{x' | \sigma(x') = \sigma(x)\} \cap \bar{V} .$$

Suppose $m = 3$. The Gaussian curvature $K(x)$ of Σ at x is

$$K(x) = [(\sigma(x) + \rho_1(x))(\sigma(x) + \rho_2(x))]^{-1} .$$

Since Σ is convex, $K(x)$ must be positive at each $x \in \bar{V}$, which implies the desired conclusion. In $\mathbb{R}^2$ the convexity of Σ (defined analogously) implies that the curvature $K(x) = [\sigma(x) + \rho_1(x)]^{-1} > 0$, and the desired conclusion follows.

Lemma 2.2. *The function* σ, *defined by* (2.1) *in two dimensions and by* (2.2) *in three dimensions, is a smooth function on* $\bar{V}$. *Furthermore,*

$$\nu = \nu(\tau(x)) = \nabla\sigma(x)$$

so that ν *is a smooth function of* x *on* $\bar{V}$. *Also* $|\nabla\sigma(x)|^2 = 1$ *on* $\bar{V}$.

Lemma 2.3. *The function* $\Delta\sigma$ *is continuous on* $\bar{V}$.

The proof of Lemma 2.2 is easy and left to the reader. Lemma 2.3 follows immediately from Lemma 2.1 and the identity

$$\Delta\sigma(x) = \Sigma_1^{m-1}[\sigma(x) + \rho_j(x)]^{-1} \quad (m=2,3) \quad .$$

3. The Basic Inequality

The starting point in obtaining our a priori estimates is the following divergence identity [1; Appendix I]

$$\text{a)} \qquad -\nabla \cdot \mathrm{Re}\,\mathfrak{L} = \mathfrak{F} + \mathfrak{Q} + \mathfrak{H} - \mathfrak{P} + \mathfrak{J}$$

where

$$\text{b)} \qquad \mathfrak{F} = -\mathrm{Re}[b \cdot \nabla\bar{u} + (i\lambda\rho + \gamma)\bar{u}]\,\frac{(\Delta u + \lambda^2 nu)}{n} ,$$

$$\text{c)} \qquad \mathfrak{Q} = -\mathrm{Re}[\bar{u}\nabla(\tfrac{\gamma}{n}) \cdot \nabla u] ,$$

$$\text{d)} \qquad \mathfrak{H} = \frac{|\nabla(\frac{\rho}{n}) \cdot \nabla u|^2}{4\omega} ,$$

$$(3.1) \qquad \text{e)} \qquad \mathfrak{P} = \omega\left|\frac{\nabla(\frac{\rho}{n}) \cdot \nabla u}{2\omega} - i\lambda u\right|^2 ,$$

$$\text{f)} \qquad \mathfrak{J} = \{\nabla\bar{u}[\nabla'(\tfrac{b}{n}) - \tfrac{1}{2}(b \cdot \nabla(\tfrac{1}{n}))I - \tfrac{\omega}{n} I]\} \cdot \nabla u ,$$

$$\text{g)} \qquad \omega = \frac{(\nabla \cdot b)}{2} - \gamma$$

and

$$\text{h)} \qquad \mathfrak{L} = [(\nabla\bar{u} \cdot \tfrac{b}{n})\nabla u - \tfrac{b}{2n}|\nabla u|^2 + \tfrac{(i\lambda\rho + \gamma)}{n}\bar{u}\nabla u + \tfrac{\lambda^2 b}{2}|u|^2] .$$

In the above definitions $\nabla'(b/n)$ is the matrix $(\partial(b^i/n)/\partial x^j)$ and I is the identity matrix. The identity (3.1) holds if u is a twice continuously differentiable complex-valued function, n, γ, ρ are real-valued, continuously differentiable functions, and b is a vector with continuously differentiable, real-valued components. Both $b(x)$ and $\gamma(x)$ must be chosen so that $\omega > 0$.

We now digress momentarily to give an alternative derivation of the identity (3.1a). This identity is a special case of the one established in Appendix I of [1], and used by Bloom there to obtain a priori estimates for solutions of elliptic equations of the form

$$(3.2)\qquad Lu = \nabla \cdot (A(x)\,\nabla u) + (a \cdot \nabla u) + \lambda^2 u = f(x)\ .$$

We present a simpler derivation of Bloom's identity here, and we specialize it to the case considered in these notes by choosing

$$A(x) = \frac{1}{n(x)}\, I\,,\quad a(x) = \frac{\nabla n(x)}{n(x)}\quad .$$

Consider the functional

$$(3.3)\qquad J(\epsilon) = \tfrac{1}{2}\int_{\Omega} [\,(\nabla \bar{u} + \epsilon\, \nabla \bar{\eta})\, E (\nabla u + \epsilon\, \nabla \eta)' - \lambda^2 n |u + \epsilon\eta|^2\,]\ ,$$

where the prime means transpose the row vector to which it is attached and where Ω is an arbitrary domain contained in the exterior region V .

If we differentiate both sides of (3.3) with respect to ϵ and set $\epsilon = 0$, we get

$$(3.4)\qquad \left.\frac{dJ}{d\epsilon}\right|_{\epsilon=0} = \mathrm{Re}\int_{\Omega} [\nabla\eta E\, \nabla \bar{u}' - \lambda n \eta \bar{u}\,]$$

$$= \mathrm{Re}\int_{\Omega} \nabla \cdot (\eta\, E\, \nabla \bar{u}')$$

$$- \mathrm{Re}\int_{\Omega} \eta\, [\nabla (E\, \nabla \bar{u}') + \lambda^2 n\, \bar{u}\,].$$

Next, setting $\eta = [\hat{b} \cdot \nabla u + (-i\lambda\hat{\rho} + \hat{\gamma})u]$ in (3.4), we obtain the identity

$$(3.5)\qquad \begin{aligned} &\mathrm{Re}\int_{\Omega} \nabla \cdot (\eta\, E\, \nabla \bar{u}') - \mathrm{Re}\int_{\Omega} \eta [\nabla (E\, \nabla \bar{u}') + \lambda^2 n\, \bar{u}\,] = \\ &\mathrm{Re}\int_{\Omega} \{ (\nabla(\hat{b} \cdot \nabla u))\, E (\nabla \bar{u})' + \nabla[(-i\lambda\hat{\rho} + \hat{\gamma})u] E\, \nabla \bar{u}' \\ &\quad - \lambda^2 n (\hat{b} \cdot \nabla u)\bar{u} - \lambda^2 n (-i\lambda\rho + \gamma)|u|^2 \}\ . \end{aligned}$$

We rewrite the integrand of the right hand side of (3.5) using the following equations

$$(3.6)\qquad \begin{aligned} \mathrm{Re}(\nabla(\hat{b} \cdot \nabla u)) E\, \nabla \bar{u}' &= \nabla u (\nabla \hat{b}') E\, \nabla \bar{u}' \\ &+ \tfrac{1}{2}\, \mathrm{Re}\, \nabla \cdot (\hat{b}\, \nabla u\, E\, \nabla \bar{u}') - \frac{(\nabla \cdot \hat{b})}{2} (\nabla u\, E\, \nabla \bar{u}') - \tfrac{1}{2}\, \mathrm{Re}\, \nabla u (\hat{b} \cdot \nabla E) \nabla \bar{u}'\ , \end{aligned}$$

$$(3.7)\qquad -\mathrm{Re}\ n(\hat{b} \cdot \nabla u)\bar{u} = -\tfrac{1}{2} \nabla \cdot (\hat{b} n |u|^2) + (\nabla \cdot (n\hat{b})) \frac{|u|^2}{2}\ ,$$

$$(3.8)\qquad -\mathrm{Re}\, i\lambda\hat{\rho}\ \nabla u E\, \nabla \bar{u}' = 0\ ,$$

and

$$(3.9)\qquad \mathrm{Re}\, \lambda^2\, n (i\lambda\hat{\rho}) |u|^2 = 0\ .$$

The last two equations are obviously true. To establish that (3.6) holds we simply note that

(3.10) $$\mathrm{Re}(\nabla(\hat{b}\cdot\nabla u'))E\nabla\bar{u}' = \nabla u(\nabla\hat{b}')E\nabla\bar{u}' + \mathrm{Re}(\hat{b}\cdot\nabla(\nabla u))E\nabla\bar{u}' ,$$

(3.11) $$\mathrm{Re}\hat{b}\cdot\nabla(\nabla u E\bar{u}') = 2\mathrm{Re}\hat{b}\cdot\nabla(\nabla u)E\nabla\bar{u}' + \nabla u\,\hat{b}\cdot\nabla E\nabla\bar{u}' ,$$

and

(3.12) $$\mathrm{Re}\,\nabla\cdot(\hat{b}\nabla u E(\nabla\bar{u})') = \mathrm{Re}\hat{b}\cdot\nabla(\nabla u E\nabla\bar{u}') + (\nabla\cdot\hat{b})(\nabla u E\nabla\bar{u}') .$$

Substituting the right hand side of (3.6) and (3.7) into (3.5) and taking account of (3.8) and (3.9), we obtain the result that

(3.13) $$\begin{gathered}\mathrm{Re}\int_\Omega \eta[\nabla\cdot(E\nabla\bar{u}') + \lambda^2 n\bar{u}] + \\ \mathrm{Re}\int_\Omega \nabla\cdot[\eta E\nabla\bar{u}' - \frac{\hat{b}}{2}(\nabla u E\nabla\bar{u}') + \frac{\lambda^2\hat{b}n}{2}|u|^2] = \\ \int_\Omega\{\nabla u[(\nabla\hat{b}' + (\hat{\gamma} - \frac{\nabla\cdot\hat{b}}{2})I)E - \hat{b}\cdot\nabla E]\nabla\bar{u}' + \\ \lambda^2(-n\hat{\gamma} + \frac{\nabla\cdot(\hat{b}n)}{2})|u|^2 + \mathrm{Re}(-i\lambda\nabla\hat{\rho} + \nabla\hat{\gamma})uE\nabla\bar{u}'\} .\end{gathered}$$

If we set $\hat{b} = n^{-1}b$, $\hat{\rho} = n^{-1}\rho$ and $\hat{\gamma} = n^{-1}\gamma$, and $E \equiv I$ then, (3.13) implies (3.1a). (To obtain the pointwise identity from the integral identity, one divides (3.13) by the volume of Ω and shrinks Ω to a point.) On the other hand, (3.13) can be rewritten as

(3.14) $$\begin{gathered}\mathrm{Re}\int_V \eta[\nabla\cdot(E\nabla\bar{u}') + (-\nabla\cdot E + a)\cdot\nabla\bar{u}' + \lambda^2 n\bar{u}] \\ + \mathrm{Re}\int_V \nabla\cdot[\eta E\nabla\bar{u}' - \frac{\hat{b}}{2}(\nabla u E\nabla\bar{u}') + \lambda^2\frac{\hat{b}n}{2}|u|^2] = \\ \int_V\{\nabla u[(\nabla\hat{b} + (\hat{\gamma} - \frac{\nabla\cdot\hat{b}}{2})I)E - \hat{b}\cdot\nabla E + (\nabla\cdot E - a)b']\nabla\bar{u}' \\ + \lambda^2(-n\hat{\gamma} + \frac{\nabla\cdot(n\hat{b})}{2})|u|^2 + \mathrm{Re}(-i\lambda\nabla\hat{\rho} + \nabla\hat{\gamma})uE\nabla\bar{u}' \\ - \mathrm{Re}(-i\lambda\hat{\rho} + \hat{\gamma})u(-\nabla\cdot E + a)\cdot\nabla\bar{u}'\} .\end{gathered}$$

This identity implies the one derived in Appendix I of [1] if we set $E \equiv A$ $n \equiv 1$, $\hat{\rho} = \rho$, $\hat{b} = a$, $\hat{\gamma} = \gamma$.

We want to bound the right-hand side of (3.1a) from above by

$$-(\nabla u\mathfrak{D})\cdot\nabla\bar{u} - \tfrac{1}{2}\mathcal{P} + A|Lu|^2 + B|u|^2 ,$$

where A and B are positive functions with $A = \mathcal{O}(r^2)$ and $B = \mathcal{O}(r^{-2})$ on $\bar{V}$ and $\mathfrak{D}$ is a positive definite matrix such that $\mathfrak{D} = \mathcal{O}(r^{-2})$ on $\bar{V}$. To this end we write

(a) $$\mathfrak{F} = \mathrm{Re}\,\Sigma_1^4\,\mathfrak{F}_i\,Lu ,$$

where

(b) $$\mathfrak{F}_1 = \frac{\rho}{n}[\frac{c\cdot\nabla u}{\omega} - i\lambda u] ,$$

(3.15) (c) $$\mathfrak{F}_2 = \left[\frac{b \cdot c}{|c|} - \frac{\rho|c|}{\omega} \right] \frac{c \cdot \nabla u}{n|c|} ,$$

(d) $$\mathfrak{F}_3 = \left[\frac{-(b \cdot c)}{n|c|} \frac{c}{|c|} + \frac{b}{n} \right] \cdot \nabla u ,$$

(e) $$\mathfrak{F}_4 = \frac{\gamma}{n} u ,$$

and

(3.16) $$Lu = \Delta u + \lambda^2 nu .$$

Next we set

(3.17) $$c = \tfrac{1}{2} \nabla \left(\frac{\rho}{n}\right) \quad \text{and} \quad \omega = 2|c|^2 .$$

Since, by definition (3.1g), $\omega = \frac{1}{2}(\nabla \cdot b) - \gamma$, the multiplier γ is now defined in terms of ρ and the vector b. It is also convenient to define two auxiliary functions μ and η on $\overline{V}$ by the equations

(3.18) $$\mu(x) = \tfrac{1}{2} + \eta(x), \; \eta(x) = \frac{\varepsilon}{3} h^2(x), \; h(x) = [\sigma(x) + \sigma_0]^{-1} ,$$

where σ is defined in Section 1 and σ_0 is some positive number such that $\sigma + \sigma_0 > 0$ on $\overline{V}$. It follows from Lemma 2.2 that μ and η are continuous functions on $\overline{V}$. We choose ε to be any positive number such that

(3.19) $$1 - \varepsilon h^2(x) > \tfrac{1}{2} \quad \text{if } \sigma \geq \sigma_0 .$$

With these definitions in mind it is easy to derive the following inequalities for the terms on the right-hand side of (3.15a):

(3.20) $$\begin{aligned} -\mathrm{Re}(\mathfrak{F}_1 L\bar{u}) &\leq \tfrac{1}{2}\mathcal{P} + A_1 |Lu|^2 , \\ -\mathrm{Re}(\mathfrak{F}_2 L\bar{u}) &\leq 2\eta \mathcal{H} + A_2 |Lu|^2 , \\ -\mathrm{Re}(\mathfrak{F}_3 L\bar{u}) &\leq -2\mu \mathcal{H} + \mu |\nabla u|^2 + A_3 |Lu|^2 , \\ -\mathrm{Re}(\mathfrak{F}_4 L\bar{u}) &\leq \left(\frac{h\gamma |u|}{2n} \right)^2 + A_4 |Lu|^2 . \end{aligned}$$

Here

$$A_1 = \left(\frac{\rho}{2n|c|} \right), \; A_2 = \frac{(\rho|c|)^2}{16\eta n^2 |c|^4} \left(\frac{(b \cdot c)}{\rho |c|^2} - 1 \right)^2 ,$$

$$A_3 = \frac{|b|^2}{4\mu n^2} , \; A_4 = h^{-2} .$$

Moreover, if we set

(3.21) $$2d = \nabla \left(\frac{\gamma}{n} \right) \quad \text{and} \quad \theta = \tfrac{1}{2}\eta > 0 ,$$

then

(3.22) $$\mathcal{Q} \leq \frac{|d|^2 |u|^2}{\theta} + \theta \, |\nabla u|^2 .$$

Using the inequalities (3.20) and (3.22), we obtain an upper bound for

$$\mathfrak{F} + \mathcal{Q} .$$

Using this estimate, the definition of ω given by (3.17) and the condition (3.19), we obtain the following basic inequality from (3.1a):

(3.23) $$-\operatorname{Re}\nabla\cdot\mathcal{L} \le -(\nabla u\mathfrak{D})\cdot\nabla\bar{u} - \tfrac{1}{2}P + A|Lu|^2 + B|u|^2 ,$$

where

(3.24) $$A = \Sigma_1^4 A_i, \quad B = \frac{\gamma^2}{4n^2 A_4} + \frac{|d|^2}{\theta} ,$$

and

(3.25) $$\mathfrak{D} = \nabla'(\tfrac{b}{n}) - \tfrac{1}{2}b\cdot\nabla(n^{-1})I - \tfrac{1}{2}(1+3\eta)I .$$

Upon integrating both sides of (3.23) over the region exterior to the obstacle ∂V and interior to a large sphere of radius R and using the Divergence Theorem to evaluate the integral of the left-hand side, we obtain the result

(3.26) $$\int_{V(R)}[(\nabla u\mathfrak{D})\cdot\nabla\bar{u} + \tfrac{1}{2}P - \int_{r=R}\frac{x}{r}\cdot\operatorname{Re}\mathcal{L} + \int_{\partial V}\nu^*\cdot\operatorname{Re}\mathcal{L} \le \int_{V(R)}[A|Lu|^2 + B|u|^2] ,$$

Here ν^* is the unit exterior normal to ∂V and $V(R) = V\cap\{x|r\le R\}$.

In the remainder of this Chapter we carry out the analysis that leads from (3.26) to estimates of the form (7.1).

The success of our further argument hinges on choosing the multipliers b and ρ so that

(3.27) $$-\int_{r=R}\frac{x}{r}\cdot\mathcal{L} \ge I_2(R) = o(1) \quad (R\to\infty) ,$$

and so that

(3.28) $$I_1 \equiv \int_{\partial V}\nu^*\cdot\operatorname{Re}\mathcal{L} \ge p_1\|u_{\nu^*}\|^2_{\partial V}$$

$$+ \text{(terms involving } u \text{ and } u_{T^*} \text{ on } \partial V) ,$$

(3.29) $$I_3 \equiv \int_V(\nabla u\mathfrak{D})\cdot\nabla\bar{u} \ge p_2\|\nabla u/r\|^2_V ,$$

and

(3.30) $$I_4 \equiv \int_V B|u|^2 \le p_3\|u/r\|^2_V ,$$

where the p_i are positive constants independent of λ .

An inequality of the form (3.28) does hold if b is chosen so that $\nu^*\cdot b$ is strictly positive on ∂V . To see this, note first that

(3.31) $$I_1 = \int_{\partial V}[(\frac{b\cdot\nu^*}{2n})|u_{\nu^*}|^2 + \operatorname{Re}\{(b\cdot\bar{u}_{T^*})u_{\nu^*}/n\} + \operatorname{Re}\{(i\lambda\rho+\gamma)\bar{u}u_{\nu^*}/n\}$$
$$+ \tfrac{1}{2}\lambda^2(\nu^*\cdot b)|u|^2 - \tfrac{1}{2}(\nu^*\cdot b)|u_{T^*}|^2] .$$

An application of the elementary inequality $ab \le \frac{1}{2}(a^2 + b^2)$ to the second and third terms in the integrand of (3.31) leads to the estimate

$$(3.32) \qquad I_1 \ge (1-2\epsilon_1)\int_{\partial V} \frac{(\nu^*\cdot b)}{2n} |u_{\nu^*}|^2 - \int_{\partial V} [F_1|u_{T^*}|^2 + F_2|u|^2],$$

where

$$F_1 = \frac{(\nu^*\cdot b)}{n} + \frac{|b|^2}{2\epsilon_1 n(\nu^*\cdot b)},$$

and

$$F_2 = \left[-\frac{\lambda^2}{2}(\nu^*\cdot b) + \frac{\lambda^2\rho^2+\gamma^2}{2\epsilon_1 n(\nu^*\cdot b)}\right].$$

The desired lower bound for I_1 follows immediately from (3.32) if we choose $\epsilon_1 > 0$ and sufficiently small, provided that both $b\cdot\nu^*$ and n are strictly positive on ∂V. The positive definiteness of $n(x)$ is one of our basic hypotheses made in the Introduction.

In Section 4 of this Chapter we shall choose ρ and b so that $b\cdot\nu^* > 0$ on ∂V, which implies (3.28) holds. We derive a lower bound for $(\nabla u\mathfrak{D})\cdot\nabla\bar{u}$ that implies (3.29) under physically reasonable hypotheses on $n(x)$. In Section 5 we derive bounds on A and B (defined in (3.24)) that imply (3.30) and

$$(3.33) \qquad \int_V A|Lu|^2 \le p_4\|rLu\|_V^2,$$

for some positive number p_4 independent of λ. In Section 6 we prove that with our choice of b and ρ made in Section 4, (3.27) holds.

We make use of (3.27) in (3.26) and take limits of both sides of the resulting inequality to obtain

$$(3.34) \qquad I_1 + \int_V [(\nabla u\mathfrak{D})\cdot\nabla\bar{u} + \tfrac{1}{2}\mathcal{P}] \le \int_V [A|Lu|^2 + B|u|^2].$$

This inequality implies the preliminary estimate (7.3) by virtue of (3.32) and the results of Sections 4-6. It follows immediately from (7.3) that a priori estimates of the form

$$(3.35) \qquad \|u_{\nu^*}\|_{\partial V}, \left\|\frac{\nabla u}{r}\right\|_V \le p\left(\left\|\frac{u}{r}\right\|_V + \|rLu\|_V + \lambda\|u\|_{\partial V} + \|u_{T^*}\|_{\partial V}\right)$$

hold, where p is a positive constant independent of u and λ. In Section 8 we use the outgoing solution $H(x,x')$ of $Lu = \delta(x,x')$ and Green's identity to derive our a priori pointwise estimate for $|u(x,\lambda)|$ using Theorem 7.1.

4. A Lower Bound for $(\nabla u\mathfrak{D})\cdot\nabla\bar{u}$

In this Section we choose b and ρ and make hypotheses on the smoothness and far-field behavior of the index of refraction $n^{\frac{1}{2}}$ sufficient to guarantee that the quadratic form $(\nabla u \mathfrak{D}) \cdot \nabla \bar{u}$ is bounded from below by a strictly positive multiple of $|\nabla u|^2$ on $\bar{V}$. Let σ_0 , $\sigma = \sigma(x)$, and $\nu = \nabla\sigma(x)$ be defined as in Section 2, and recall the definition (3.18) of h . We further recall from Section 2 that $\sigma(x)$, $\tau(x)$ are the coordinates of $x \in \bar{V}$ under the transformation (2.1) in $\mathbb{R}^2$ and (2.2) in $\mathbb{R}^3$.

HYPOTHESES: <u>There exist constants</u> $p > 2$, $n_0 > 0$, $c_i > 0 (i = 1,2)$, <u>and</u> $\sigma_1 \geq \sigma_0$ <u>such that in both the case of</u> $\mathbb{R}^2$ <u>and the case of</u> $\mathbb{R}^3$

$$(4.1) \qquad \begin{aligned} &(i) \quad n \in C^2(V) \cap C^1(\bar{V}), \\ &(ii) \quad n(x) \geq n_0 > 0 \ (x \in \bar{V}) ; \end{aligned}$$

$$(4.2) \qquad \begin{aligned} &(i) \quad |1 - n^{-1}(x)| \leq c_2 h^p(x) \ (x \in \bar{V}, \ \sigma(x) \geq \sigma_2), \\ &(ii) \quad |n_\sigma| \equiv |\nu \cdot \nabla n| \leq c_1 h^p(x) \quad (x \in \bar{V}, \ \sigma(x) \geq \sigma_2). \end{aligned}$$

<u>Further, there exists a positive number</u> c_0 <u>such that for all</u> $x \in \bar{V}$

$$(4.3) \qquad n + \tfrac{1}{2} h^{-1} (\nu \cdot \nabla n) \geq c_0 .$$

Note that hypothesis (4.3) restricts the magnitude of $n_\sigma = \nu \cdot \nabla n$ only on subsets of $\bar{V}$ where $n < 0$. The hypotheses (4.2) and (4.3) are physically reasonable. Radical changes in n relative to its magnitude can produce concentration of energy. The hypotheses we make ensure that energy flows out to infinity at a reasonable rate. This is consistent with our local-energy norm interpretation of our a priori L_2 - estimates, which we mentioned in Chapter 0 .

We next introduce an auxiliary function $\Gamma(x)$ to be used in the definition of the multiplier $b(x)$.

<u>Definition</u> 4.1. Γ is the $C^1(\bar{V})$-function with piecewise continuous second derivatives given by

$$(4.4) \qquad \Gamma(x) = \begin{cases} 1 + \varepsilon'^{-1} & \text{if } \sigma(x) \leq \sigma_2 \\ \varepsilon'^{-1}[1 + \varepsilon' - (k(x) - 1)^2] & \text{if } \sigma(x) > \sigma_2 , \end{cases}$$

where $k(x) = (\sigma_2 + \sigma_0) \, h(x)$.

In this definition ε' is a positive number, which we shall later assume to be suitably small. We now define our multipliers b and ρ .

<u>Definition</u> 4.2. The multiplier b is the $C^1(\bar{V})$-vector function with piecewise continuous second derivatives defined by

$$(4.5) \qquad b(x) \equiv n(x)\Gamma(x)h^{-1}(x)\nabla\sigma(x) ,$$

and ρ is the $C^1(\bar{V})$-function defined by

$$(4.6) \qquad \rho(x) \equiv n(x)[h^{-1}(x) + 2(\sigma_2 + \sigma_0)(\varepsilon')^{-1} + \varepsilon h(x)] .$$

In (4.6) ε is the positive number chosen in (3.18) and ε' is the positive number occuring in (4.4).

The following lemma plays a key role in the proof of Proposition 4.1 below.

Lemma 4.4. *If we define* $\hat{b}(x) = \nabla\sigma(x)/h(x)$, *then the matrix* $\nabla'\hat{b} \equiv (\hat{b}^i_j)$ *has the property*

$$\nabla'\hat{b} \geq I \quad (x \in \overline{V}) ; \tag{4.7}$$

that is, for each complex m-vector ψ, $(\psi\nabla'\hat{b}) \cdot \psi \geq \psi \cdot \psi$ *on* $\overline{V}$.

Proof. We give the proof only in the case of $\mathbb{R}^3$. It is easy to specialize the argument to $\mathbb{R}^2$. We first observe that

$$\hat{b} = \alpha + \nu\sigma_0 ,$$

where $\alpha = \nu\sigma = (\nabla\sigma)\sigma$ is the vector multiplier used in [2]. In a typical coordinate zone (see Section 2)

$$\hat{b}^i_j = \alpha^i_j + \sigma_0\sigma_{ij} \quad (i,j = 1, 2, 3) , \tag{4.8}$$

where the subscripts denote differentiation with respect to x^j and, as computed in [2; Eqn. (8.7)],

$$\alpha^i_j = \delta_{ij} - \Sigma^2_{\ell=1} \frac{\rho_\ell X^{0i}_{\sigma_\ell} X^{0j}_{\sigma_\ell}}{(\sigma+\rho_\ell)|X^0_{\sigma_\ell}|^2} . \tag{4.9}$$

On the other hand, since $\sigma_i = X^i_\sigma$, in a typical coordinate zone

$$\sigma_{ij} = (X^i_\sigma)_j = \Sigma^2_{\ell=1} \frac{X^{0i}_{\sigma_\ell} X^{0j}_{\sigma_\ell}}{(\sigma+\rho_\ell)|X^0_{\sigma_\ell}|^2} . \tag{4.10}$$

Substituting for α^i_j and σ_{ij} from (4.9) and (4.10) in (4.8), we find that in a typical coordinate zone

$$\nabla'\hat{b} = I + \Sigma^2_{\ell=1} \frac{(\sigma_0 - \rho_\ell)(X^0_{\sigma_\ell})' X^0_{\sigma_\ell}}{(\sigma+\rho_\ell)|X^0_{\sigma_\ell}|^2} , \tag{4.11}$$

where $(X^0_{\sigma_\ell})'$ is the transpose of $X^0_{\sigma_\ell}$ and

$$(X^0_{\sigma_\ell})' X^0_{\sigma_\ell} = (X^{0i}_{\sigma_\ell} X^{0j}_{\sigma_\ell}) .$$

By Lemma 2.1 both $\sigma + \rho_1$ and $\sigma + \rho_2$ are positive on $\overline{V}$. We now further restrict our choice of σ_0 so that both

$$\sigma + \sigma_0 > \text{ and } \sigma_0 - \rho_\ell > 0 \ (\ell = 1,2) \text{ in } \overline{V} . \tag{4.12}$$

Having thus chosen σ_0, we see that since

$$(\overline{\psi}[(X^0_{\sigma_\ell})'X^0_{\sigma_\ell}]) \cdot \psi = |\Sigma^3_{i=1} \psi^i X^{0i}_{\sigma_\ell}|^2 \geq 0$$

on $\bar{V}$, the desired inequality (4.7) holds on $\bar{V}$. This completes the proof of Lemma 4.1.

We now state and prove the main result of this Section, which holds in $\mathbb{R}^m (m=2,3)$.

Proposition 4.1. Suppose that σ_0 satisfies (4.12), b and ρ are chosen according to Definition 4.2, ϵ and ϵ' are sufficiently small, and the hypotheses (4.1) - (4.3) are satisfied. Then the matrix $\mathfrak{D}$, which is defined by (3.25), (3.18), and the choice of b and σ_0, is strictly positive definite on $\bar{V}$. In particular, there exists a positive constant C independent of λ such that on $\bar{V}$

$$(\nabla u \mathfrak{D}) \cdot \nabla \bar{u} \geq Ch^2(x)|\nabla u|^2 \quad . \tag{4.13}$$

Our Proof of Proposition 4.1 is considerably simpler than the proof of the analogous result in [1; Appendices II and III]. Our choice of ρ is simpler, and we avoid the patching argument used by Bloom.

Proof. Using the definition of b, we rewrite $\mathfrak{D}$ as

$$\mathfrak{D} = \Sigma_1^4 \mathfrak{D}_i , \tag{4.14}$$

where

$$\mathfrak{D}_1 = \Gamma(\nabla'\hat{b} + \frac{\hat{b}\cdot\nabla n}{2n} I), \qquad \mathfrak{D}_2 = (\nabla\Gamma)'\hat{b} ,$$

$$\mathfrak{D}_3 = -\frac{1}{2n}|\nabla(\frac{\rho}{n})|^2 I, \qquad \mathfrak{D}_4 = -\tfrac{1}{2}(1+3\eta)I .$$

We consider two cases.

Case I: $x \in S_I$, where $S_I = \{x|\sigma(x) \leq \sigma_2\} \cap \bar{V}$.

In this case Γ is constant so that $\mathfrak{D}_2 = 0$. Furthermore, the constant value of Γ can be made arbitrarily large by taking ϵ' sufficiently small. Therefore, if

$$\nabla'\hat{b} + \frac{\hat{b}\cdot\nabla n}{2n} I > 0 , \tag{4.15}$$

then $\mathfrak{D}_1$ can be made as large as we please by taking ϵ' small enough. On the other hand, both $\mathfrak{D}_3$ and $\mathfrak{D}_4$ are independent of ϵ' and bounded on S_I since S_I is compact. Thus if (4.15) holds, then $\mathfrak{D}$ is positive definite on S_I. But by Lemma 4.1

$$\nabla'\hat{b} + \frac{(\hat{b}\cdot\nabla n)}{2n} I \geq [1 + \frac{(\nu\cdot\nabla n)}{2nh(x)}] I . \tag{4.16}$$

By our hypothesis (4.3) the right-hand side of (4.16) is positive definite; hence (4.13) holds, and $\mathfrak{D}$ is positive definite on S_I. (Our hypothesis (4.3) should be compared to the conditions II-2 of [1].)

Finally, since S_I is compact, it follows that on S_I

$$(\nabla u \mathfrak{D}) \cdot \nabla \bar{u} \geq C_I h^2(x)|\nabla u|^2 , \tag{4.17}$$

where

$$C_I = \operatorname*{Min}_{S_I} [h^{-2}(x) (\operatorname*{Min}_{|\psi|=1} (\psi \mathfrak{D}) \cdot \overline{\Psi})] .$$

Case II. $x \in S_{II}$, where $S_{II} = \{x | \sigma(x) \geq \sigma_2\} \cap \overline{V}$.

Our approach is direct: we compute and estimate the terms in (4.14). The far-field behavior of n now comes into play. The hypotheses (4.2) imply that

$$\mathfrak{D}_1 \geq \Gamma[1 - C_1 h^p(x)(1 + C_2 h^p(x))]I . \tag{4.18}$$

The choice of σ_2 does not affect our proof. We choose it so large that the function in square brackets in (4.18) is positive.

Estimates of $\mathfrak{D}_2$ and $\mathfrak{D}_3$ are easy to obtain. We find by straight-forward computation that

$$\begin{aligned} \mathfrak{D}_2 &= \frac{-2}{\epsilon'} k(x)[1 - k(x)] (\nabla\sigma)'\nabla\sigma , \\ \mathfrak{D}_3 &= -\tfrac{1}{2}[1 - \epsilon h^2(x)]^2 + \mathcal{O}(h^p(x)) . \end{aligned} \tag{4.19}$$

In estimating $\mathfrak{D}_3$ we have again used the hypothesis (4.2).

Recalling the definition (3.5) of η, we see that

$$\mathfrak{D}_4 = -\tfrac{1}{2}[1 + \epsilon h^2(x)] . \tag{4.20}$$

We now use the estimates (4.18)-(4.20) to conclude that for all $x \in S_{II}$

$$\begin{aligned} (\nabla u \mathfrak{D}) \cdot \nabla \overline{u} \geq{} & \Gamma[1 + \mathcal{O}(h^p(x))]|\nabla u|^2 - \frac{2}{\epsilon'} k(x)[1 - k(x)]|\nu \cdot \nabla u|^2 \\ & + [-1 + \tfrac{1}{2}\epsilon h^2(x)[1 - \epsilon h^2(x)] + \mathcal{O}(h^p(x))]|\nabla u|^2 . \end{aligned} \tag{4.21}$$

But $\Gamma = 1 + (\Gamma - 1)$, and

$$\Gamma - 1 = \frac{k(x)}{\epsilon'} [2 - k(x)] . \tag{4.22}$$

Making use of (4.22) in (4.21), we find that

$$\begin{aligned} (\nabla u \mathfrak{D}) \cdot \nabla \overline{u} \geq{} & \{[1 + \mathcal{O}(h^p(x))] + \frac{1}{\epsilon'} k(x)[2 - k(x)][1 + \mathcal{O}(h^p(x))] \\ & -1 + \frac{\epsilon h^2(x)}{2} [1 - \epsilon h^2(x)] + \mathcal{O}(h^p(x))\} |\nabla u|^2 \\ & - \frac{2}{\epsilon'} k(x)[1 - k(x)] |\nu \cdot \nabla u|^2 . \end{aligned} \tag{4.23}$$

Since $-|\nu \cdot \nabla u| > -|\nabla u|$, we deduce from (4.23) that on S_{II}

$$\begin{aligned} (\nabla u \mathfrak{D}) \cdot \nabla \overline{u} &\geq \{(\epsilon')^{-1}(\sigma_2 + \sigma_0)^2 + \tfrac{1}{2}\epsilon [1 - \epsilon h^2(x)] + \mathcal{O}(h^{p-2}(x))\} h^2(x) |\nabla u|^2 \\ &\overset{d}{=} q(x) h^2(x) |\nabla u|^2 . \end{aligned} \tag{4.24}$$

We now choose σ_2 so large that it satisfies our previous conditions and

$$\operatorname*{Inf}_{\sigma \geq \sigma_2} q(x) > 0 .$$

With this choice of σ_2 it follows from (4.24) that

(4.25) $$(\nabla u \mathfrak{D}) \cdot \nabla \bar{u} \geq C_{II} h^2(x) |\nabla u|^2 ,$$

where

$$C_{II} = \operatorname*{Min}_{\sigma \geq \sigma_2} [h^{-2}(x) (\operatorname*{Min}_{|\psi|=1} (\psi \mathfrak{D}) \cdot \bar{\psi})].$$

Finally, on the basis of (4.17) and (4.25) we conclude that Proposition (4.1) holds with $C \stackrel{d}{=} \mathrm{Min}(C_I , C_{II})$.

5. Far-field Behavior of Coefficients of the $|Lu|^2$ and $|u|^2$ - terms

In this Section we establish the behavior as $\sigma \to \infty$ of the coefficients of $|Lu|^2$ and $|u|^2$ in our basic inequality (3.23). We use the choices of b and ρ made in the last Section.

Several straightforward calculations using the hypotheses (4.1) and (4.2) lead to the conclusion that, as $\sigma \to \infty$,

(5.1) $$A_2, A_3 = \mathcal{O}(\sigma^2) , \qquad A_1 + A_4 = \mathcal{O}(\sigma^2) ,$$

where the A_i are defined after (3.20). In view of (5.1) the coefficient A of $|Lu|^2$ in (3.23) grows no faster than a constant multiple of σ^2 as $\sigma \to \infty$, and

(5.2) $$K_3 = \operatorname*{Sup}_{V} [h^2(x)A] < \infty .$$

Next, we examine the behavior of the coefficient B of $|u|^2$ in (3.23) as $\sigma \to \infty$. We use the assumptions (4.1) and (4.2) and we also assume:

<u>Hypotheses</u>. <u>For</u> <u>all</u> $x \in \bar{V}$

(5.3)
$$\text{(i)} \quad |\nabla n| \leq C_3 h^2(x) ,$$
$$\text{(ii)} \quad |n_{\sigma\sigma}| \leq C_4 h^3(x) ,$$
$$\text{(iii)} \quad |n_{\sigma\tau_i}| \leq C_5 h^2(x) \qquad (i = 1, m - 1)$$

<u>in</u> $\mathbb{R}^m$ $(m = 2,3)$. These conditions are implied by hypotheses (H) - (vii) and (viii) in the Introduction.

The calculations necessary to obtain the estimates to follow are tedious and straightforward for the most part. Therefore we omit most of them. All the big-oh estimates in the remainder of this Section hold for $x \in \bar{V}$.

First, it follows from the definitions of ω, ρ, and Γ that

$$(5.4)\qquad \begin{aligned} &(a)\quad \frac{\rho}{n} = \mathcal{O}(h^{-1}(x)),\\ &(b)\quad 2\omega = |\mathrm{grad}(\rho/n)|^2 = (\rho/n)_\sigma^2 = \mathcal{O}(1),\\ &(c)\quad \mathrm{grad}\,\omega = (\rho/n)_\sigma(\rho/n)_{\sigma\sigma}\nu = \mathcal{O}(h^2(x)),\end{aligned}$$

and

$$(d)\quad \Gamma = \mathcal{O}(1),\ \Gamma_\sigma = \mathcal{O}(h^2(x)),\ \Gamma_{\sigma\sigma} = \mathcal{O}(h^3(x)).$$

Furthermore

$$(5.5)\qquad \nabla\sigma = \sum_{\ell=1}^{m-1} \frac{1}{(\sigma+\rho_\ell)} = \mathcal{O}(\sigma^{-1})\ (m=2,3);$$

and since $|\mathrm{grad}\,\rho_\ell| = \mathcal{O}(1)$,

$$(5.6)\qquad \mathrm{grad}(\nabla\sigma) = -\sum_{\ell=1}^{m-1} \frac{(\nu+\mathrm{grad}\,\rho_\ell)}{(\sigma+\rho_\ell)^2} = \mathcal{O}(\sigma^{-2}).$$

A straightforward calculation using (5.4d) and (5.5) gives the result

$$\begin{aligned}\nabla\cdot b &= n_\sigma\Gamma h^{-1} + n[h^{-1}\Gamma_\sigma + \Gamma] + n\Gamma h^{-1}\Delta\sigma\\ &= n_\sigma\mathcal{O}(\sigma) + n\mathcal{O}(1).\end{aligned}$$

Thus by the hypotheses (4.2),

$$(5.7)\qquad \nabla\cdot b = \mathcal{O}(1).$$

This result together with (5.4b) implies that

$$(5.8)\qquad \gamma = \tfrac{1}{2}\nabla\cdot b - \omega = \mathcal{O}(1).$$

An immediate consequence of the last estimate and our hypotheses on n is that

$$(5.9)\qquad \frac{\gamma h^2}{4n} = \mathcal{O}(\sigma^{-2}).$$

We next estimate d^2/θ, recalling that $d = \mathrm{grad}(\gamma/n)$. This calculation is rather tedious since second derivatives of b, and hence n, are involved through grad γ. In estimating grad(div b) we use the results (5.4d), (5.5) and (5.6) to estimate Γ and the derivatives of Γ and σ, while we use the hypotheses (4.1), (4.2) and (5.3) to estimate n and its derivatives. We find that

$$(5.10)\qquad \nabla(\nabla\cdot b) = \mathcal{O}(\sigma^{-2}).$$

This result together with (5.4c) implies that

$$(5.11)\qquad \nabla\gamma = \mathcal{O}(\sigma^{-2}).$$

Finally, using the results (5.9) and (5.8) in the definition of d, we obtain the estimate

$$(5.12)\qquad d^2/\theta = \mathcal{O}(\sigma^{-2})$$

since $\theta = \mathcal{O}(\sigma^{-2})$ by (3.18).

In view of the estimates (5.9) and (5.12), we conclude that B grows no faster than a constant multiple of σ^{-2} as $\sigma \to \infty$; that is,

(5.13) $$K_4 = \sup_V [h^{-2}(x)B] < \infty .$$

6. The Radiation Integral

Once cannot expect that Problem P, stated in the Introduction, will have a unique solution unless an outgoing radiation condition such as (1.3) is imposed. It follows that somewhere in a proof of a priori estimates for solutions of Problem P (estimates that imply uniqueness) the radiation condition must be used. It plays its role in this Section. Our goal is to derive the result (3.27), namely to prove that

(6.1) $$-\int_{r=R} \frac{x}{r} \cdot \operatorname{Re} \mathcal{L} \geq o(1) \quad (R \to \infty) .$$

This result is used to deduce (3.34) from (3.26). We first rewrite the integrand in (6.1) as a quadratic form in u, u_T, and $\mathcal{D}_1 u$, where

(6.2) $$\mathcal{D}_1 u \overset{d}{=} u_r - i\lambda u + \frac{(m-1)}{2r} u$$

and u_T is the component of grad u lying in the plane perpendicular to the position vector x on the sphere $|x| = r$. To do this we write

$$\nabla u = u_r \frac{x}{r} + u_T T, \quad u_r = \mathcal{D}_1 u + i\lambda u - \frac{(m-1)}{2r} u$$

and

$$b^r = \frac{x \cdot b}{r}, \quad b^T = T \cdot b .$$

The result is:

(6.3)
$$\begin{aligned} -\int_{r=R} \frac{x}{r} \cdot \operatorname{Re} \mathcal{L} = \int_{r=R} \Big\{ & \frac{b^r}{2n} [|u_T|^2 - |\mathcal{D}_1 u|^2] - \operatorname{Re}[\frac{b^T}{n} \bar{u}_T \mathcal{D}_1 u] \\ & + \frac{|u|^2}{n} \{\lambda^2[(\rho - b^r) + \tfrac{1}{2} b^r (1-n)] + \frac{\gamma(m-1)}{2r} - \tfrac{1}{2} b^r (\frac{m-1}{2r})^2\} \\ & - \operatorname{Re} \{[i\lambda(\rho - b^r) + (\gamma - \frac{m-1}{2r} b^r)] \frac{\bar{u}}{n} \mathcal{D}_1 u\} \\ & - \operatorname{Re} \{(i\lambda - \frac{m-1}{2r}) b^T \frac{\bar{u}_T}{n} u\} \Big\} . \end{aligned}$$

We use the inequality $|ab| \leq \tfrac{1}{2} [c|a|^2 + c^{-1}|b|^2]$ to estimate the cross-product terms involving u_T in (6.3). We choose $c = \tfrac{1}{2}$ so that the resultant $|u_T|^2$-terms exactly cancel the term $b^r|u_T|/2n$ in (6.3). Note that b^r is positive for sufficiently large r since

$$b^r = r + \mathcal{O}(1) \quad (r \to \infty),$$

(see (6.7) below). We next use the inequality $|ab| \leq \tfrac{1}{2}[r|a|^2 + r^{-1}|b|^2]$, to estimate the $u\mathcal{D}_1 u$-terms in (6.3). We thus obtain

(6.4)
$$-\int_{r=R} \frac{x}{r}\cdot \operatorname{Re}\mathcal{L} \geq \hat{I}_2(R) + J(R)\ ,$$

where

(6.5)
$$\hat{I}_2(R) = \int_{r=R} \frac{r|\mathcal{D}_1 u|^2}{n}\left\{-\tfrac{1}{2}\frac{b^r}{r} - \frac{|b^T|^2}{rb^r} - 1\right\}$$

and

(6.6)
$$J(R) = \int_{r=R} \frac{|u|^2}{n}\Bigg\{\lambda^2\left[(\rho - b^r) - \frac{|b^T|^2}{b^r} + \tfrac{1}{2}b^r(1-n) - \frac{(\rho - b^r)^2}{2r}\right] + \frac{\gamma(m-1)}{2r}$$
$$- \tfrac{1}{2}b^r\left(\frac{m-1}{2r}\right)^2 - \left(\frac{m-1}{2r}\right)^2\frac{|b^T|^2}{b^r} - \left(\gamma - \frac{(m-1)b^r}{2r}\right)^2\frac{1}{2r}\Bigg\}\ .$$

We use the local coordinate transformation (2.1) or (2.2) to estimate the terms within curly brackets in these last two integrals in each coordinate zone. Of course, we employ our various hypotheses on n. The results are that as $r \to \infty$,

(6.7)

(a) $\nu_r = 1 + \tfrac{1}{2}r^{-2}(\nu\cdot X^0 - |X^0|^2) + \mathcal{O}(r^{-4})$,

(b) $\sigma = r - \nu\cdot X^0 + (2r)^{-1}(\nu\cdot X^0 - |X^0|^2) + \mathcal{O}(r^{-3})$,

(c) $b^r = r + [\dfrac{2(\sigma_2+\sigma_0)}{\varepsilon'} + \sigma_0 - \nu\cdot X^0] + \mathcal{O}(r^{-1})$,

(d) $\rho - b^r = [\dfrac{(\sigma_2+\sigma_0)^2}{\varepsilon'} + \varepsilon - \tfrac{1}{2}(\nu\cdot X^0 - |X^0|^2)]r^{-1} + \mathcal{O}(r^{1-p}) + \mathcal{O}(r^{-2})$,

(e) $1 - n = \mathcal{O}(r^{-p})$,

(g) $\dfrac{|b^T|^2}{b^r} = -r^{-1}(\nu\cdot X^0 - |X^0|^2) + \mathcal{O}(r^{-2})$,

(h) $\gamma - \dfrac{m-1}{2r}b^r = -\Sigma_1^{m-1}\rho_\ell r^{-1} + \mathcal{O}(r^{-1})$.

It follows that the term multiplying $\dfrac{r|\mathcal{D}_1 u|^2}{n}$ in (6.5) is $\mathcal{O}(1)$ as $r \to \infty$. Therefore, by the Radiation Condition (1.3),

(6.8)
$$\hat{I}_2(R) = o(1) \quad (R \to \infty)\ .$$

Next we observe that the coefficient of λ^2 in the integrand of (6.6) can be made positive. We accomplish this by first choosing ε' small enough to make the coefficient of the $\mathcal{O}(r^{-1})$ part of

$$\rho - b^r - \frac{|b^T|^2}{b^r}$$

positive and then choosing R so large that this $\mathcal{O}(r^{-1})$ term dominates, say $R > R_0$. Then we can choose λ so large, say $\lambda \geq \lambda_0$, that the integrand in (6.6) is non-negative if $R \geq R_0$, since the remaining terms not involving λ in the curly brackets are $\varepsilon'^{-1}\,\mathcal{O}(r^{-1})$. Thus

(6.9) $$J(R) \geq 0 \quad \text{for} \quad R \geq R_0 \quad \text{and} \quad \lambda \geq \lambda_0 .$$

The results (6.8) and (6.9) imply (6.1) if $\lambda \geq \lambda_0$.

We close this Section with a lemma concerning radiation integrals that will be useful a little later on.

Lemma 6.1. If $u \in C^1(V)$ and u satisfies the radiation condition

$$\lim_{R\to\infty} \int_{r=R} r|\mathcal{D}_1 u|^2 = 0 ,$$

then for each $\delta > 0$

(6.10) $$\lim_{R\to\infty} \int_{r=R} r^{-1-\delta}|u|^2 = 0 .$$

Proof. It is a direct consequence of the divergence theorem that in R^m

$$\int_{r=R} \frac{|u|^2}{r} - \int_{\partial V} \frac{\nu^* \cdot x}{r^2}|u|^2 \equiv \int_{V(R)} \operatorname{div}\left(\frac{x|u|^2}{r^2}\right) = \int_{V(R)} \frac{2\operatorname{Re}\bar{u}\,u_r}{r} + (m-2)\int_{V(R)} \frac{|u|^2}{r^2} ,$$

where $V(R)$ is the intersection of V with a large ball of radius R . Recall that

$$u_r = \mathcal{D}_1 u + i\lambda u - \frac{(m-1)}{2r}\, u .$$

Therefore,

$$\int_{r=R} \frac{|u|^2}{r} = \int_{\partial V} \frac{\nu^* \cdot x}{r^2}|u|^2 + \int_{V(R)} \frac{2\operatorname{Re}\bar{u}\,\mathcal{D}_1 u}{r} - \int_{V(R)} \frac{|u|^2}{r^2} .$$

It follows that

$$\int_{r=R} \frac{|u|^2}{r} \leq \int_{\partial V} \frac{\nu^* \cdot x}{r^2}|u|^2 + \int_{V(R)} |\mathcal{D}_1 u|^2 .$$

The integral over ∂V is bounded by our assumptions on u . Thus we may rewrite the preceeding inequality as

$$\int_{r=R} \frac{|u|^2}{r} \leq \text{const.} + \int_{V(R)} |\mathcal{D}_1 u|^2 .$$

By virtue of the radiation condition,

$$\int_{V(R)} |\mathcal{D}_1 u|^2 = \mathcal{O}(\ln R) \quad (R \to \infty) .$$

Therefore

$$\int_{r=R} \frac{|u|^2}{r} \leq \mathcal{O}(\ln R) \quad (R \to \infty) .$$

We now divide both sides of the last relation by R^δ to obtain the desired conclusion.

7. A Priori Estimates in Weighted L_2 - norms

The results obtained in Sections 3-6 together with an auxiliary estimate for

$\|u/r\|_V$, to be derived in this Section, imply the following theorem.

Theorem 7.1. Suppose u is a solution of Problem P, with f replaced by $g(x,\lambda)$, that lies in $C^2(V)\cap C^1(\bar{V})$. Assume that V can be illuminated from the exterior. Finally assume that the hypotheses (4.1)-(4.3) and (5.3) each hold. Then if λ is sufficiently large, there exist positive constants Γ_1 and Γ_2, independent of λ and u, such that

$$(7.1)\quad \begin{aligned} &\text{(a)}\ \|u_{\nu*}\|_{\partial V},\ \|(\text{grad } u)/r\|_V \le \Gamma_1\,[\|u\|_{\partial V}+\|u_{T*}\|_{\partial V}+\|rg\|_V],\\ &\text{and}\\ &\text{(b)}\ \|u/r\|_V \le \frac{\Gamma_2}{\lambda}\,[\lambda\|u\|_{\partial V}+\|u_{T*}\|_{\partial V}+\|rg\|_V]\,. \end{aligned}$$

Here $\|\cdot\|_S$ is the L_2-norm over the set S. The inequalities of Theorem (7.1) imply the following corollaries.

Corollary 7.2. Under the hypotheses of Theorem (7.1), if λ is sufficiently large, there exists a positive constant Γ_3, independent of λ and u such that

$$E_{R_1}(e^{-i\lambda t}u) \overset{d}{=} [\lambda^2\|u\|_{V(R_1)} + \|\text{grad } u\|^2_{V(R_1)}]/2$$

$$\le \Gamma_3[\lambda^2\|u\|^2_{\partial V}+\|u_{T*}\|^2_{\partial V}+\|rg\|^2_V]\,,$$

where $E_{R_1}(e^{-i\lambda t}u)$ is the portion of the energy of $ue^{-i\lambda t}$ contained in $V(R_1) = V\cap\{x|\ |x|\le R_1\}$ and

$$\Gamma_3 = [\min_{V(R_1)} r^{-2}][\Gamma_1^2+\Gamma_2^2]\,.$$

Corollary 7.3. Under the hypotheses of Theorem 7.1, if λ is sufficiently large, then Problem P has a unique solution in the class $C^2(V)\cap C^1(\bar{V})$.

The remainder of this Section is devoted to proving Theorem 7.1. From the result obtained in Section 6 that (3.33) holds for functions which satisfy the Radiation Condition (1.3) and from (3.26), we conclude that (3.34) holds for the solution u of Problem P. We now observe that the vector multiplier b has been chosen so that on ∂V, $\nu*\cdot b>0$ (the auxilliary function Γ is a positive constant on ∂V). The multiplier ρ is strictly positive on ∂V. Therefore it follows from (3.32) that

$$(7.2)\qquad I_1 \ge K_2\|u_{\nu*}\|^2_{\partial V} - K_5\|u_{T*}\|^2_{\partial V} - \lambda^2K_6\ \|u\|^2_{\partial V}\,,$$

where

$$K_2 = (1-2\epsilon_1)[\inf_{\partial V}(\frac{\nu*\cdot b}{n})] > 0\ (0<\epsilon_1<\tfrac{1}{2})\,,$$

$$K_5 = \sup_{\partial V} F_1 < \infty\,,\quad\text{and}\quad K_6 = \lambda^{-2}\sup_{\partial V} F_2 < \infty\,.$$

Making use of (7.2), Proposition 4.1, and the results (5.2) and (5.13) in (3.34), we conclude that if u is the solution of Problem P, then for λ sufficiently large,

$$(7.3)\qquad \begin{aligned} &K_1\|h\nabla u\|_V^2 + \tfrac{1}{2}\int_V \wp + K_2\|u_{\nu*}\|_{\partial V}^2 \\ &\le K_3\|hu\|_V^2 + K_4\|h^{-1}g\|_V^2 + K_5\|u_{T*}\|_{\partial V}^2 + \lambda^2 K_6\|u\|_{\partial V}^2 , \end{aligned}$$

where

$$(7.4)\qquad K_1 = \min(C_I, C_{II}) > 0$$

and K_3 and K_4 are defined by (5.2) and (5.13).
Note that the K_i $(i = 1, \ldots 5)$ are independent of λ and $K_6 = \mathcal{O}(1)$ as $\lambda \to \infty$.

It follows from the ray equation $x = \nu\sigma + X^0(\tau)$ that in each coordinate zone

$$r = [\sigma^2 + 2(\nu \cdot X^0(\tau))\sigma + |X^0(\tau)|^2]^{\frac{1}{2}} .$$

Since $\nu \cdot X^0 > 0$, $r \ge \sigma$ on $\bar{V}$ wherever $\sigma(x) > 0$. Thus there exist constants ω_i $(i = 1, 2)$ such that

$$(7.5)\qquad \|h^{-1}g\|_V^2 \le \omega_1\|rg\|_V^2 , \quad \text{and} \quad \|hw\|_V^2 \ge \omega_2 \|w/r\|_V^2 ,$$

where w is either u or ∇u. The inequalities (7.5) and (7.3) imply (3.35) for λ sufficiently large, say $\lambda \ge \lambda_0$, with

$$(7.6)\qquad p = \frac{\max(\omega_2 K_3, \omega_1 K_4, K_5, K_6|_{\lambda=\lambda_0})}{\min(\omega_2 K_1, K_2)}$$

It still appears that we are far from our goal to obtain a priori estimates for $\|u/r\|_V$, $\|\nabla u/r\|_V$, and $\|u_{\nu*}\|_{\partial V}$. For while the integrals on the left-hand side of (3.35) or (7.3) are the unknown quantities we wish to estimate a priori, they are bounded from above in (3.35) or (7.3) by a linear combination of given quantities and the unknown quantity $\|u/r\|_V$ or $\|hu\|_V$. However we shall now demonstrate that the latter quantity can be bounded from above by the sum of small multiples of the quantities we desire to estimate in (7.1a) and small multiples of known quantities (see (7.12 below). These multiples can be made as small as we please by choosing λ sufficiently large. By using this bound on $\|hu\|_V^2$ in (7.3) we obtain an inequality that immediately yields (7.1a). We then use the estimates for the quantities on the left-hand side of (7.1a) in the "small multiples" estimate (7.12) for $\|hu\|_V^2$ to obtain an inequality that immediately yields (7.1b).

We begin to carry out this program with the identity

$$(7.7)\qquad \begin{aligned} \nabla \cdot (h^2\bar{u}\nabla u) &= h^2|\nabla u|^2 - 2h^3\bar{u}(\nabla\sigma \cdot \nabla u) \\ &\quad - \lambda^2 h^2 n |u|^2 + (\bar{u} L u)h^2 . \end{aligned}$$

This identity holds for solutions of Problem P. We integrate it over $V(R)$ and use the divergence theorem. The result after letting $R \to \infty$ is:

$$(7.8)\quad \begin{aligned} -\int_{\partial V} h^2 \bar{u}\, u_{\nu*} &= \int_V [h^2\{\bar{u}g + |\nabla u|^2\} - 2h^3 \bar{u}\,(\nabla\sigma \cdot \nabla u)] \\ &- \lambda^2 \int_V h^2 n|u|^2 - \lim_{R\to\infty} \int_{r=R} h^2 \bar{u}\, u_r \,. \end{aligned}$$

We next show that

$$(7.9)\quad \lim_{R\to\infty} \int_{r=R} h^2 \bar{u}\, u_r = 0 \,.$$

To prove (7.9) we observe that

$$(7.10)\quad \begin{aligned} \Big|\int_{r=R} h^2 \bar{u}\, u_r\Big| &= \Big|\int_{r=R} h^2 \bar{u}\, \mathcal{D}_1 u + \int_{r=R} (i\lambda - \frac{(m-1)}{2r})\, h^2|u|^2\Big| \\ &\le \text{const.} \int_{r=R} |\mathcal{D}_1 u|^2 + \text{const.}\ (\tfrac{1}{2} + |\lambda| + \frac{m-1}{2r}) \int_{r=R} \frac{|u|^2}{r^2} \,. \end{aligned}$$

But by Lemma 6.1 and the Radiation Condition (1.3), the right-hand side of the last inequality converges to 0 as $R \to \infty$. This implies (7.9).

We next estimate the various terms that remain in (7.8) with the help of the inequality $ab \le \frac{1}{2}[c|a|^2 + c^{-1}|b|^2]$. The result is

$$(7.11)\quad \begin{aligned} (1 - \frac{2}{\lambda^2}) \int_V h^2 n|u|^2 &\le \frac{\lambda^{-2}}{2} \int_{\partial V} [\varepsilon_2 |u_{\nu*}|^2 + h^4|u|^2] \\ &+ \lambda^{-2} \int_V h^2 [\frac{|g|^2}{4n} + (1 + \frac{h^2}{n})\, |\nabla u|^2] \,. \end{aligned}$$

If λ is so large that $[1 - 2\lambda^{-2}] > 0$ and $n(x) \ge n_0$ on $\bar{V}$ (see (4.1 - ii)), then (7.11) implies the "small multiples" estimate

$$(7.12)\quad \begin{aligned} \|hu\|_V^2 &\le \lambda^{-2} (1 - \frac{2}{\lambda^2})^{-1} [D_1 \|u_{\nu*}\|_{\partial V}^2 + D_2 \|h\nabla u\|_V^2 \\ &+ D_3 \|u\|_{\partial V}^2 + D_4 \|hg\|_V^2] \,, \end{aligned}$$

where

$$D_1 = \tfrac{1}{2} n_0^{-1},\quad D_2 = n_0^{-1} [1 + \operatorname*{Sup}_V (h^2 n^{-1})]$$

$$D_3 = \tfrac{1}{2} n_0^{-1} \operatorname*{Sup}_V h^4,\quad D_4 = \frac{1}{4n_0^2} \,.$$

Using (7.12) to estimate the unknown term on the right-hand side of (7.3), we obtain the result that for all $\lambda \ge \lambda_0$

$$(7.13)\quad \begin{aligned} \|u_{\nu*}\|_{\partial V}^2 + \|h\nabla u\|_V^2 &+ \tfrac{1}{2} \|\omega^{\frac{1}{2}} | \frac{c \cdot \nabla u}{\omega} - i\lambda u|\, \|_V^2 \\ &\le \Lambda_1 [\lambda^2 \|u\|_{\partial V}^2 + \|u_{T*}\|_{\partial V}^2 + \|h^{-1} g\|_V^2] \,. \end{aligned}$$

Here

$$(7.14)\quad \Lambda_1 = \left\{ \text{Min}\left[1,\ K_1 - \frac{D_2K_3}{\lambda_0^2}(1 - \frac{2}{\lambda_0^2})^{-1},\ K_2 - \frac{D_1K_3}{\lambda_0^2}(1 - \frac{2}{\lambda_0^2})^{-1}\right]\right\}^{-1} \cdot \text{Max}\left[K_5,\ K_6 + \frac{K_3D_3}{\lambda_0^4}(1 - \frac{2}{\lambda_0})^{-1},\ K_4 + \frac{K_3D_4(1 - 2\lambda_0^{-2})^{-1}}{\lambda_0^2\,(\underset{V}{\text{Inf}}\ h^{-4})}\right],$$

and the number λ_0 is so large that all the quantities involved in the definition of Λ_1 are positive, $[1 - 2\lambda_0^{-2}]$ is positive, and (7.3) holds for $\lambda \geq \lambda_0$.

Next we employ (7.13) and (7.5) in (7.12) to find that for $\lambda \geq \lambda_0$

$$(7.15)\quad \|u/r\|_V^2 \leq \lambda^{-2}(1 - \frac{2}{\lambda_0^2})^{-1}\ \Lambda_2[\lambda^2\|u\|_{\partial V}^2 + \|u_{T*}\|_{\partial V} + \|h^{-1}g\|_V^2\ ],$$

where

$$(7.16)\quad \Lambda_2 = \frac{1}{\omega_2}\text{Max}[\ \Lambda_1D_1 + \Lambda_1D_2 + D_3,\ (\Lambda_1D_1 + \Lambda_1D_2),\ (\Lambda_1D_1 + \Lambda_1D_2) + \frac{D_4}{\epsilon_1}[\underset{V}{\text{Inf}}\ h^{-4}]^{-1}],$$

and λ_0 is so large that Λ_1 and Λ_2 are both positive.

The estimates of Theorem 7.1 are direct consequences of (7.13), (7.15) and (7.5). Corollary 7.2 is a simple consequence of Theorem 7.1. Corollary 7.3 follows from the pointwise a priori estimate for $|u(x, \lambda)|$ that we establish in the next Section using Theorem 7.1.

8. An A Priori Estimate for $|u(x, \lambda)|$

Our final objective in this Chapter is to obtain an a priori estimate for $|u(x, \lambda)|$ which holds uniformly in x on $\overline{V}$ for $\lambda \geq \lambda_0$. We first derive an upper bound for $|u|$ in terms of

$$\|rg\|_V,\ \max_{x\in\partial V}|u_0(x)|,\ \|u/r\|_V,\ \text{and}\ \|u_{\nu*}\|_{\partial V}.$$

Then, making use of the inequalities (7.1) of Theorem 7.1, to estimate the last two quantities, we obtain the desired pointwise estimate for $|u(x, \lambda)|$.

Let $H(x,x')$ be the (fundamental) solution of

$$(8.1)\quad \Delta H + \lambda^2 H = \delta(x,x')$$

that satisfies the Radiation Condition (1.3); namely, let

(8.2)

$$\text{(a)}\quad H(x,x') = \frac{e^{i\lambda|x-x'|}}{|x-x'|}\quad (x,\ x' \in \mathbb{R}^3),$$

$$\text{(b)}\quad H(x,x') = \frac{i}{4}H_0^1(\lambda|x-x'|)\quad (x,\ x' \in \mathbb{R}^2),$$

where $H_0^1(z)$ is the Hankel function of first kind of order zero.

As usual, we begin with an identity to which we shall apply the Divergence

Theorem:

$$\nabla\cdot(u\nabla H) - \nabla\cdot(H\nabla u) = -H(\Delta u + \lambda^2 n(x')u) + \lambda^2 n(x')Hu \tag{8.3}$$
$$+ u(\Delta H + \lambda^2 H) - \lambda^2 uH .$$

Here u is the solution of Problem P with f replaced by g and H is as just defined above. The variables of differentiation in (8.3) are the x' variables. We integrate (8.3) over the region $V(R)$, which is the intersection of V with a large ball of radius R. The result after applying the divergence theorem is:

$$u(x,\lambda) = \int_{V(R)} H(x,x')\{g(x',\lambda) - \lambda^2[n(x')-1]u(x')\}dx'$$
$$+ \int_{\partial V} \{H(x,x')u_{\nu*}(x',\lambda) - u(x',\lambda)H_{\nu*}(x,x')\}dS(x') \tag{8.4}$$
$$+ \int_{|x'|=R} (uH_r - Hu_r)\, dS(x') .$$

Since

$$\int_{|x'|=R} (uH_r - Hu_r) \equiv \int_{|x'|=R} (u\mathcal{D}_1 H - H\mathcal{D}_1 u) ,$$

it is possible to conclude that

$$\lim_{R\to\infty} \int_{|x'|=R} (uH_r - Hu_r) = 0 . \tag{8.5}$$

To see this note first that for $\delta > 0$

$$\left|\int_{|x'|=R} u\mathcal{D}_1 H\right| \le \left[\int_{|x'|=R} \frac{|u|^2}{r^{1+\delta}}\right]^{\frac{1}{2}} \left[\int_{|x'|=R} r^{1+\delta}|\mathcal{D}_1 H|^2\right]^{\frac{1}{2}} . \tag{8.6}$$

The first integral on the right-hand side of (8.6) has the limit 0 as $R\to\infty$ by Lemma 6.1. The second integral is $\mathcal{O}(R^{\delta-3})$ for $m = 2$ or 3 by the properties of H. Thus choosing $\delta = 1$, we conclude that the left-hand side of (8.6) has limit zero as $R\to\infty$. Similarly we find that

$$\left|\int_{|x'|=R} H\mathcal{D}_1 u\right| = \mathcal{O}(R^{-\frac{1}{2}}) \quad (R\to\infty) .$$

These results imply (8.5).

We let $R\to\infty$ in (8.4) and use (8.5). With a little care we conclude from the resultant identity that [4; Lemma 3]

$$|u(x,\lambda)| \le r^{(1-m)/2}\left\{\left[\max_{x\in\bar V}\|r^{(m-1)/2}(x)r^{-1}(\cdot)H(x,\cdot)\|_V\right]\|rg\|_V\right.$$
$$+ \lambda^2\left[\max_{x\in\bar V} r^2\left|\frac{1-n}{n}\right|\right]\left[\max_{x\in\bar V}\|r^{-1}(\cdot)r^{(m-1)/2}(x)H(x,\cdot)\|_V\right]\|u/r\|_V \tag{8.7}$$
$$+ \lambda\left[\max_{x\in\bar V}\int_{\partial V}\lambda^{-1}r^{(m-1)/2}(x)|H_{\nu*}(x,\cdot)|\right]\max_{x\in\partial V}|u_0|$$
$$\left.+ \left[\max_{x\in\bar V}\|r^{(m-1)/2}(x)H(x,\cdot)\|_{\partial V}\right]\|u_{\nu*}\|_{\partial V}\right\} .$$

It can be shown that the factors involving H in (8.7) are bounded by $C\lambda^{-(3-m)/2}$ $(m = 2,3)$, where C is a constant independent of λ; see [1], for example. Consequently, it follows from (8.7) that if λ is sufficiently large,

$$(8.8)\qquad |u(x,\lambda)| \le C'\lambda^{-(3-m)/2} r^{(1-m)/2} \{\|rg\|_V + \lambda^2\|\tfrac{u}{r}\|_V + \lambda \max_{x\in\partial V} |u_0| + \|u_{\nu*}\|_{\partial V}\} ,$$

where C' is some positive number independent of x and λ.

Finally, using Theorem 1 to estimate the terms that are a priori unknown on the right-hand side of (8.8), we obtain:

Theorem 8.1. *If the hypotheses of Theorem 7.1 hold and λ is sufficiently large ($\lambda \ge \lambda_0$), then there exists a positive constant Γ_3, independent of λ and x, such that for $\lambda \ge \lambda_0$ and all x in $\overline{V}$*

$$(8.9)\qquad |u(x,\lambda)| \le \Gamma_3 \lambda^{(1+m)/2} r^{(1-m)/2} \{\|rg\|_V + \|u_{T*}\|_{\partial V} + \lambda \max_{x\in\partial V} |u_0|\} ,$$

where u is the solution of Problem P.

References for Chapter 1

1. Bloom, C.O., Estimates for solutions of reduced hyperbolic equations of the second order with a large parameter. J. Math. Anal. Appl. 44(1973) 310-332.
2. Bloom, C.O. and Kazarinoff, N.D., Local energy decay for a class of nonstar-shaped bodies, Archive for Rat. Mech. and Anal., 55 (1974), 73-85.
3. Feldman, E.A., The geometry of immersions. II. Bull. Amer. Math. Soc. 70(1964), 600-607.
4. Morawetz, C.S. and Ludwig, D., An inequality for the reduced wave operator and the justification of geometrical optics. Comm. Pure Appl. Math. 21(1968), 187-203.
5. Morawetz, C.S., Decay for solutions of the Dirichlet exterior problem for the wave equation, Comm. Pure Appl. Math. 28(1975), 229-264.
6. Strauss, W.A., Dispersion of waves vanishing on the boundary of an exterior domain. Comm. Pure Appl. Math. 28(1975), 265-278.

CHAPTER 2

GLOBAL EXISTENCE, SMOOTHNESS, AND NONFOCUSSING OF OPTICAL PATHS IN A REFRACTIVE MEDIUM

1. Introduction

In this Chapter we consider the following problems: Let ∂V be a scattering obstacle (∂V is an $(m-1)$-sphere smoothly embedded in $\mathbb{R}^m$, $m = 2$ or 3) in an optical medium with variable index of refraction $n^{\frac{1}{2}}$.

(1) Find conditions on ∂V and $n^{\frac{1}{2}}$ sufficient to guarantee that the optical paths (rays) emanating normally from ∂V fill the exterior region V without intersecting (focussing). (Under these conditions we say that ∂V is convex relative to $n^{\frac{1}{2}}$; see Definition 2.3 below.)

(2) Find sufficient conditions for the rays emanating from a proper connected subset S of ∂V to fill some unbounded, connected region T of the exterior of ∂V without intersecting each other. Under these conditions we say that S is locally convex relative to $n^{\frac{1}{2}}$; see Definition 2.2 below and Fig. 1a. If S is locally convex relative to $n^{\frac{1}{2}}$ and the rays emanating normally from S do not intersect $\partial V - S$, then we say that S is a locally convex patch of ∂V; see Fig. 1b.

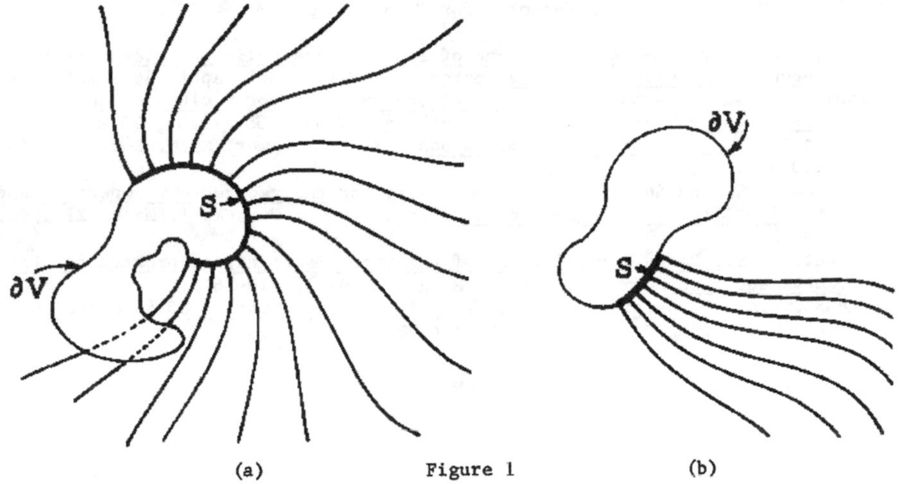

(a) Figure 1 (b)

(3) Consider the coordinate systems on connected subsets of $\bar{V} = V \cup \partial V$ defined by curves, normally incident on ∂V and extending to infinity, and the surfaces orthogonal to them. Among these coordinate systems characterize those which have the property that the curves normally incident on ∂V are optical paths (geodesics in the Riemannian metric given by $ds = n^{\frac{1}{2}}|dx|$) for some index of refraction $n^{\frac{1}{2}}$.

(4) Furthermore, find subsets of the coordinate systems characterized in (2) that define smooth indices of refraction $n^{\frac{1}{2}}$ on V or $T \cup S$ such that $n(x) \to 1$

at an appropriate rate as $|x| \to \infty$.

As we pointed out in Chapter 0, the study of families of rays in relation to indices of refraction is fundamental to the analysis of any scattering problem in a optical (or acoustical) medium, and hence the above questions are of significance. In Chapter 3 the results of this Chapter are applied to prove that a function

$$u_N \equiv \sum_{k=1}^{K} e^{i\lambda \chi_k(x)} \Sigma_{j=0}^{N+1} A^{kj}(x)\lambda^{-j} + \Sigma_{j=2}^{N+1} B^{j-2}(x)\lambda^{-j}$$

is, with appropriate choices of the functions χ_k, A^{kj} and B^j, a rigorous asymptotic approximation as $\lambda \to \infty$ to the solution of the radiating body problem

$$\text{(P)} \qquad \begin{aligned} &\Delta u + \lambda^2 n(x) u = f(x) \qquad (x \in V), \\ &u|_{\partial V} = u_0(x), \\ &\lim_{R\to\infty} \int_{r=R} r\left|u_r - i\lambda u - \frac{(m-1)}{2r} u\right|^2 = 0 \quad (r = |x|), \end{aligned}$$

if ∂V is an obstacle that consists of K disjoint patches $S_1, \ldots, S_K$ $(K \geq 1)$ that are each locally convex relative to $n^{\frac{1}{2}}$ and which are smoothly joined together. Note that the $j = 0$ term in the double sum and the $j = 2$ term in the single sum yield the approximation to the field given by the classical theory of geometrical optics; see [4] and [5]. The index of refraction $n^{\frac{1}{2}}$ can be considered as given, or as defined by given ray coordinate systems; see Corollaries 5.2 and 5.3 and Definitions 2.1 - 2.3.

We give answers to the questions posed above in Lemma 2.4, Theorem 2.5, Theorem 2.7, Proposition 4.1, Theorem 5.1, Corollaries 5.2 and 5.3, and Theorems 6.1 and 8.1 below. These are the main results in this Chapter. D.S. Jones [3] has obtained related, but less general results in the case of $\mathbb{R}^2$, but the above relationships between ray systems and indices of refraction $n^{\frac{1}{2}}$ have not been previously studied in $\mathbb{R}^3$ under general hypotheses on $n^{\frac{1}{2}}$.

If we assume, as we shall do, that $n(x) \to 1$ as $|x| \to \infty$, then it is reasonable to expect that each optical path emanating normally from a given surface is asymptotic to a straight line as $|x| \to \infty$. Hence, given a field $\hat{\mathfrak{F}}$ of straight lines [5, p. 108] emanating normally from a smooth closed convex surface (curve) in $\mathbb{R}^3$ ($\mathbb{R}^2$), there should exist some (largest) neighborhood $\mathfrak{n}$ of infinity where the optical paths $\mathfrak{F}$ asymptotic to the lines in $\hat{\mathfrak{F}}$ do not intersect. The rays of $\mathfrak{F}$ should form a field on $\mathfrak{n}$; the optical paths in $\mathfrak{F}$ and the surfaces orthogonal to them should define families of (ray) coordinate systems on a finite number of overlapping tubular subsets T_i of $\mathfrak{n}$ such that $\cup_i T_i \supseteq \mathfrak{n}$; see Corollary 5.3. (The orthogonal surfaces (curves) of $\mathfrak{F}$ are convex relative to $n^{\frac{1}{2}}$.) See Fig. 2.

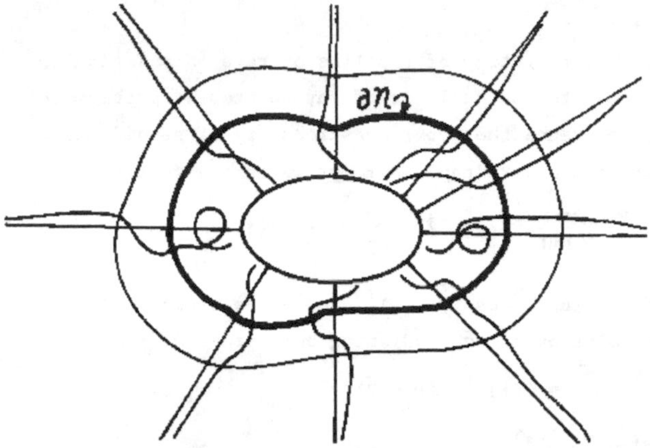

Figure 2

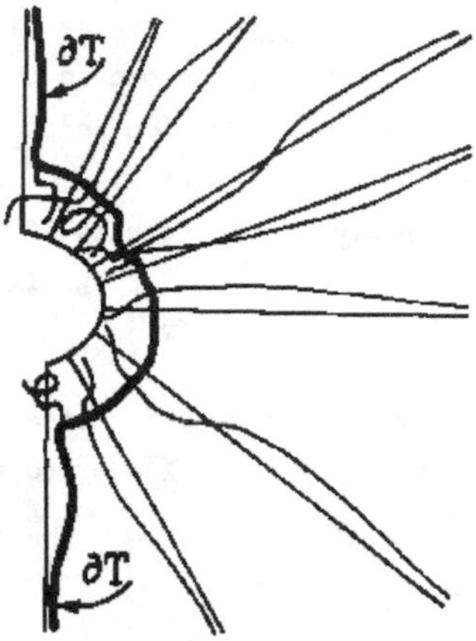

Figure 3

We prove under physically reasonable hypotheses on $n^{\frac{1}{2}}$ that the above conjectures are correct. Our proof shows that if meticulous track is kept of all constants, then a specific numerical estimate for the size of the neighborhood $\mathfrak{n}$ can be found in terms of n and the rate at which $n \to 1$ as $|x| \to \infty$.

Under the same conditions on $n^{\frac{1}{2}}$, we also prove in this Chapter that if a field $\hat{\mathfrak{F}}$ of straight lines emanating normally from a subset of a smooth <u>locally convex</u> surface (curve) in $\mathbb{R}^3$ $(\mathbb{R}^2)$ is given, there is a tubular region T extending to infinity where the optical paths asymptotic to the lines in $\hat{\mathfrak{F}}$ do not intersect, that is where these optical paths form a field $\mathfrak{F}$; see Fig. 3. This result implies Corollary 5.3. (The optical paths in $\mathfrak{F}$ and the surfaces (curves) orthogonal to them define a <u>(ray) coordinate system on</u> $T \cup S$. The orthogonal surfaces (curves) <u>of</u> $\mathfrak{F}$ <u>are locally convex relative</u> to $n^{\frac{1}{2}}$.) Our analysis uses classical differential geometry and the fixed point theorem for contracting maps.

The setting of this Chapter is that of optics, but our results can be interpreted in other settings as well. For example, the differential equations for optical paths in a medium with a variable index of refraction $n^{\frac{1}{2}}$ can be interpreted as the differential equations for the trajectory of a particle in a force field determined by $n^{\frac{1}{2}}$; see [4]. Thus our results are global theorems asserting existence, nonfocussing, and smoothness of trajectories of a class of autonomous dynamical systems. Optical paths in a medium with a variable index of refraction $n^{\frac{1}{2}}$ may also be interpreted as geodesics in a Riemannian geometry with metric tensor nI. Consequently, our theorems also assert global existence, nonfocussing, and smoothness of geodesics in such Riemannian geometries. Of course, another implication of the results of this Chapter is the existence on connected unbounded regions of $\mathbb{R}^m$ $(m = 2,3)$ of smooth solutions of the eikonal equation $|\nabla\sigma|^2 = n(x)$.

We hope that the analysis of this Chapter opens the way to analogous studies for more general scattering problems, involving both inhomogeneous and anisotropic media, e.g. problems in which there are multiple reflections, caustics, or diffraction phenomena. In every case it is necessary to establish the existence of the physically relevant geometrical-optical paths; and if a purely formal asymptotic solution of a scattering problem can be found, the smoothness and asymptotic properties of these optical paths are the key to proving that the partial sums of the formal asymptotic series solution are rigorous approximations to the exact solution.

This Chapter is organized as follows. In Section 2 we first give sufficient conditions that a coordinate system defined by a family of curves normally incident on a connected subset of ∂V are geodesics in an appropriate Riemannian metric; see Lemma 2.4 and Theorem 2.5. Also in Section 2 (Theorem 2.7 and Lemma 2.6) and in Sections 5 and 6 we prove that fields of geodesics in the Riemannian metric defined by a given index of refraction are ray coordinate systems (see Definition 2.1 below). These results complete the solution to the third problem posed in the first paragraph of this Introduction.

Sections 3-7 are devoted to solving the first and second problems posed in the first paragraph of this Introduction. In Section 3 we prove an existence and uniqueness theorem for a general class of nonlinear Volterra integral equations. We apply this theorem in Section 4 to the integral equation (2.18) shown in Section 2 to be satisfied by a geodesic of a given Riemannian metric $ds = n^{\frac{1}{2}}|dx|$, that is asymptotic to a straight line (of a given field of lines) as $|x| \to \infty$. The conclusion stated in Proposition 4.1, together with the results of Section 6 imply that (i) given a sufficiently smooth index of refraction $n^{\frac{1}{2}}(x)$ defined on $\mathbb{R}^3$ ($\mathbb{R}^2$) that approachs 1 sufficiently fast as $|x| \to \infty$ (see Hypotheses (2.20)-(2.23)), and (ii) an $(m-1)$-parameter field $\hat{\mathfrak{F}}$ of straight lines emanating normally from a surface (curve) that is convex or locally convex, then there exists a unique geodesic (solution of 2.18) that is asymptotic to each member of $\hat{\mathfrak{F}}$.

In Section 5 we consider the family of geodesics corresponding to a given $n^{\frac{1}{2}}$, and a given field of straight lines $\hat{\mathfrak{F}}$ emanating normally from a locally convex surface (curve) $\hat{S}$ that is a connected proper subset of a smoothly embedded 3(2) sphere. We prove that such a family of geodesics forms a field $\mathfrak{F}$ on an unbounded connected subset T of the region in $\mathbb{R}^m$ covered by the lines in $\hat{\mathfrak{F}}$; see Theorem 5.1 and Corollary 5.2. (Every point in T is required to lie outside some sufficiently large neighborhood of $x = 0$.) Furthermore, as a consequence of the results stated in Theorem 5.1, we obtain the result that if $\hat{S}$ is a closed surface (curve), then the rays corresponding to $\hat{\mathfrak{F}}$ form a field $\mathfrak{F}$ on a neighborhood $\mathfrak{n}$ of infinity in $\mathbb{R}^3$ ($\mathbb{R}^2$) (see Corollary 5.3).

In view of Proposition 4.1, Theorem 5.1, Corollary 5.2, (6.2), and (6.3), the elements of $\mathfrak{F}$ satisfy conditions (2.1)-(2.6) of Definition (2.1) if $\hat{S}$ is not closed. Moreover, by virtue of (6.1a), (6.1b), and the hypotheses made on $n(x)$ (see (2.20)-(2.23)), the reasoning leading to Theorem 2.5 can be applied to the geodesics in $\mathfrak{F}$ to prove that these geodesics satisfy conditions (2.7) and (2.8) of Definition 2.1. It then follows that the optical paths in $\mathfrak{F}$, and the wave fronts orthogonal to them define an <u>orthogonal ray coordinate system</u> on the subset $\overline{T}$ and, a fortiori, that the eikonal equation (2.13) is satisfied, as well as the ray equation (2.12).

The fourth problem posed in the first paragraph of this Introduction is dealt with at the end of Section 7. There we give conditions on the ray coordinate systems sufficient to assure that (i) n and a specified number of derivatives of n are continuous, and (ii) that $n(x) - 1$ and a specified number of its derivatives tend to zero at an algebraic rate with respect to r as $r = |x| \to \infty$; cf. (2.20)-(2.23).

2. Ray Coordinate Systems and Convexity Relative to $n^{\frac{1}{2}}(x)$

Consider an (m-1)-parameter family $\mathfrak{F}$ of smooth curves

$$\{x \,|\, x = X(\sigma, \tau),\ \tau = \text{const.},\ \tau \in D',\ \sigma \geq 0\}$$

in $\mathbb{R}^m$ (m = 2, 3), where $\tau = (\tau^1, \tau^{m-1})$ and D' is a closed connected subset of the rectangular domain $\tau_1 \leq \tau < \tau_2$.

Suppose that:

(2.1) $\{x \,|\, x = X(0, \tau),\ \tau \in D'\}$

is a smooth embedding of a closed, connected, proper subset of an (m-1)-sphere (which is topologically an (m-1)-cell). We naturally call such a set a patch on the embedded (m-1)-sphere.

(2.2) $X_\sigma(0, \tau) \cdot X_{\tau^i}(0, \tau) = 0 \quad (i = 1, m-1)$ for all $\tau \in D'$.

(2.3) Each curve in $\mathfrak{F}$ extends to infinity without intersecting itself or any other curve in the family $\mathfrak{F}$.

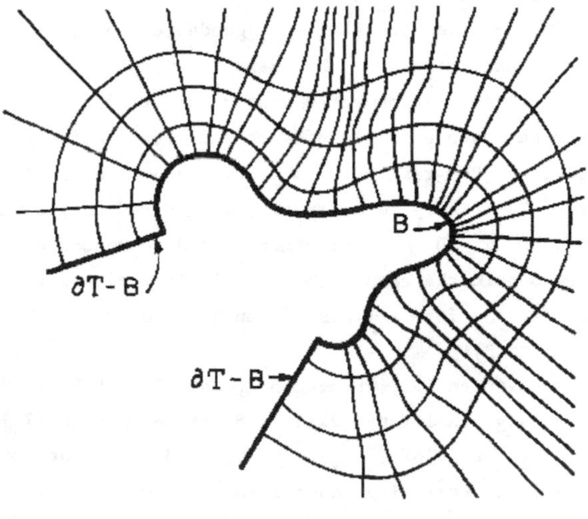

Figure 4

(2.4) The curves in $\mathfrak{F}$ cover the region T bounded by the surface (curve) $B = \{x | x = X(0, \tau), \tau \in D'\}$ and the surface (curves) $\{x | x = X(\sigma, \tau), (\sigma, \tau) \in [0, \infty) \times \partial D'\}$ in $\mathbb{R}^3$ $(\mathbb{R}^2)$; see Fig. 4.

(2.5) The mapping $(\sigma, \tau) \to X(\sigma, \tau)$ is a one-to-one mapping of $\mathcal{S} = [0, \infty) \times D'$ onto $\overline{T} = T \cup \partial T$.

Under the assumptions (2.1)-(2.5), the curves

$$\{x | x = X(\sigma, \tau), \sigma \geq 0, \tau = \text{const.}\},$$

$$\{x | x = X(\sigma, \tau), \sigma = \text{const.}, \tau^1 = \text{const.}\},$$

$$\{x | x = X(\sigma, \tau), \sigma = \text{const.}, \tau^{m-1} = \text{const.}\},$$

where $\tau \in D'$ and $\sigma \geq 0$, define a coordinate system on $\overline{T}$ that is a simple covering of $\overline{T}$.

Remark. The hypothesis (2.5) implies (2.3).

We shall demonstrate below that if, in addition,

(2.6) the Jacobian $J = \partial X/\partial(\sigma, \tau)$ is positive for all $(\sigma, \tau) \in \mathcal{S}$; and

(2.7) $X_\sigma(\sigma, \tau) \cdot X_{\tau i}(\sigma, \tau) = 0$ $(i = 1, m-1)$ for all $(\sigma, \tau) \in \mathcal{S}$,

then the curves in $\mathfrak{F}$ are geodesics of the Riemannian metric

$$ds = n^{\frac{1}{2}}|dx|,$$

where

(2.8)
$$n^{\frac{1}{2}}(x) = |X_\sigma(\sigma(x), \tau(x))|^{-1} = |\nabla\sigma(x)|$$

and $\sigma = \sigma(x)$, $\tau = \tau(x)$ $(x \in \overline{T})$ is the transformation inverse to $x = X(\sigma, \tau)$, $(\sigma, \tau) \in \mathcal{S}$.

These curves are optical paths (of least time) in the inhomogeneous medium governed by the index $n^{\frac{1}{2}}$, and they satisfy the classical first order differential equations for optical paths; see Lemma 2.4 below.

Definition 2.1. An (m-1)-parameter family of curves $\mathfrak{F}$ is an orthogonal ray coordinate system corresponding to a given index of refraction $n^{\frac{1}{2}}$ on a region $\overline{T} \subset \mathbb{R}^m$ (in brief, a ray system) if and only if the conditions (2.1)-(2.7) and the relationship (2.8) hold.

Definition 2.2. Let Σ be a closed, connected, proper subset of a surface (curve) that is a smooth embedding of an $(m-1)$-sphere in $\mathbb{R}^m$ $(m = 2, 3)$. We call Σ a patch. We say that a patch Σ is locally convex relative to $n^{\frac{1}{2}}$ if and only if it is an orthogonal surface (curve) (that is a wave front) of an orthogonal ray coordinate system $\mathfrak{F}$ that defines $n^{\frac{1}{2}}$ (by (2 8)).

In particular, B (or any connected subset of B) where B is the base of the tubular region defined in (2.4), is locally convex relative to $n^{\frac{1}{2}}$ if the conditions (2.1)-(2.8) hold; see Fig. 5. Definition 2.2 applies to the case $\Sigma = S$ and $S \subseteq \partial V$.

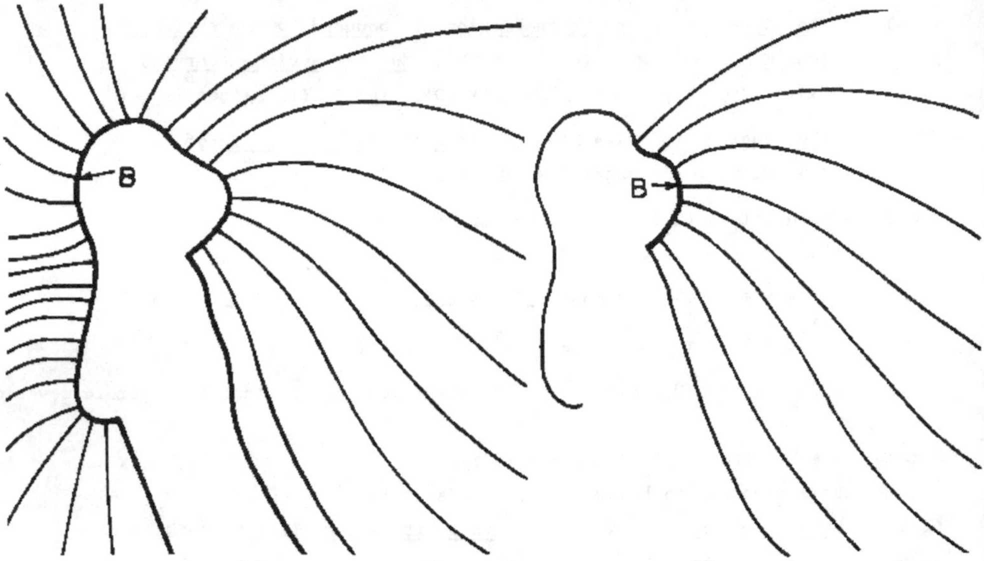

(a) Figure 5 (b)

<u>Definition</u> 2.3. If Σ is a smooth embedding of a 2(1)-sphere in $\mathbb{R}^3$ ($\mathbb{R}^2$), then Σ is <u>convex</u> <u>relative</u> <u>to</u> $n^{\frac{1}{2}}$ if and only if it is an orthogonal surface (curve) (that is, a wave front of a field of geodesics of the Riemannian metric defined by $ds = n^{\frac{1}{2}}|dx|$ that cover ext Σ).

Note that a locally convex (relative to $n^{\frac{1}{2}}$) patch Σ is not necessarily a subset of a closed surface (curve) that is convex relative to $n^{\frac{1}{2}}$. This is obvious if $n^{\frac{1}{2}}(x) \equiv 1$; see Fig. 6a. The situation for $n^{\frac{1}{2}}(x) \not\equiv 1$ is depicted in Fig. 6b.

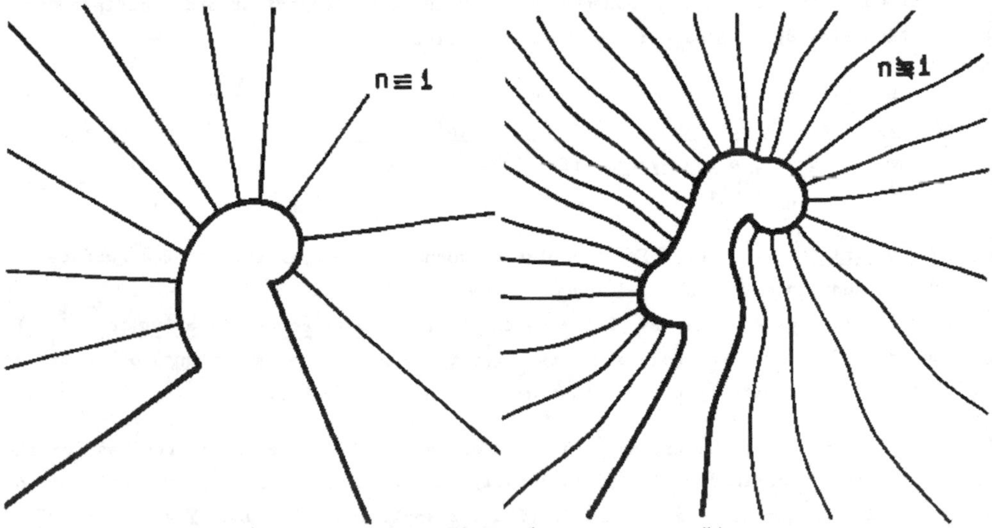

(a) Figure 6 (b)

As we demonstrate below in Corollary 5.2, this will be the case provided Σ is the union of a finite number of patches $S_1, S_2, \ldots, S_K$ that are locally convex relative to $n_1^{\frac{1}{2}}, n_2^{\frac{1}{2}}, \ldots, n_K^{\frac{1}{2}}$, respectively, (where $n_j^{\frac{1}{2}}(x)$ is defined by (2.8) on $\overline{T}_j$) and provided:

(i) the rays in $\mathfrak{F}_j$ emanating normally from $S_j \cap S_k$ belong to $\mathfrak{F}_k (j \neq k)$,

(ii) no ray in $\mathfrak{F}_j$ emanating normally from $S_j - S_k$ intersects any geodesic in $\mathfrak{F}_k$ $(j \neq k)$,

(iii) $\operatorname{ext} \Sigma \cup \Sigma = \cup_{j=1}^{j=K} \overline{T}_j$

(iv) $n^{\frac{1}{2}}(x) = n_j^{\frac{1}{2}}(x) \quad (x \in \overline{T}_j)$,

is a continuous function on $\overline{\operatorname{ext} \Sigma}$; see Fig. 7. In particular, the above definition applies to the case where $\Sigma = \partial V$.

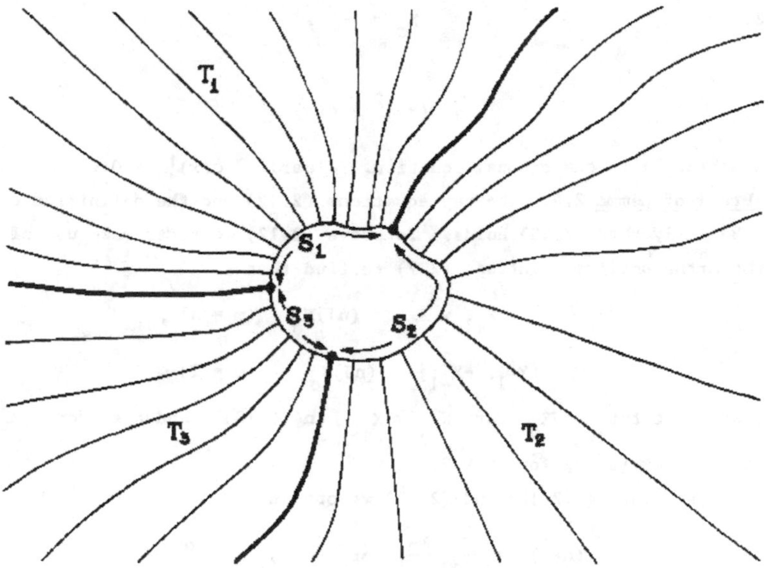

Figure 7

Suppose that $X(\sigma, \tau)$ is a ray system, as defined above. Let (σ, τ) with $\sigma = \sigma(x)$, $\tau = \tau(x)$ be the transformation inverse to the mapping given by $x = X(\sigma, \tau)$. Then the differential identity

(2.9) $$dX^i = X^i_\sigma \, d\sigma + \Sigma_{j=1}^{m-1} X^i_{\tau^j} \, d\tau^j \quad (i = 1, \ldots, m)$$

holds. From (2.5), the relations

$$(2.10)\qquad \text{(a)}\quad \nabla\sigma = \frac{X_{\tau 1} \times X_{\tau 2}}{J} \quad (m = 3)\ ,$$

$$\text{(b)}\quad \nabla\sigma = \frac{[X^2_{\tau 1}, -X^1_{\tau 1}]}{J} = (m = 2)$$

follow, where the Jacobian

$$(2.11)\qquad \text{(a)}\quad J = X_\sigma \cdot (X_{\tau 1} \times X_{\tau 2}) \quad (m = 3)\ ,$$

$$\text{(b)}\quad J = X_\sigma \cdot [X^2_{\tau 1}, -X^1_{\tau 2}] \quad (m = 2)$$

is positive.

Using (2.8) and the orthogonality conditions $X_{\tau i} \cdot X_\sigma = 0$ $(i = 1, m-1)$, it is easy to prove the following lemma.

Lemma 2.4. If X defines a ray system and we define $n^{\frac{1}{2}}(x) = |X_\sigma|^{-1}$, then

$$(2.12)\qquad X_\sigma = \frac{\nabla\sigma}{n}\ ,$$

and

$$(2.13)\qquad |\nabla\sigma(x)|^2 = n(x)\ .$$

Equation (2.13) is the eikonal equation. Clearly, $\sigma(x)|_B = 0$.

Proof of Lemma 2.4. The ray equations (2.12) and the definition of $n^{\frac{1}{2}}$ provided by (2.8) imply that (2.13) holds. To prove (2.12) we first make use of (2.8), (2.11), and the orthogonality relations (2.7) to find that

$$X_{\tau 1} \times X_{\tau 2} = (nJ)X_\sigma \quad (m = 3)\ ,$$

$$[X^2_{\tau 1}, -X^1_{\tau 1}] = (nJ)X_\sigma \quad (m = 2)\ .$$

Next, we substitute nJX_σ for $X_{\tau 1} \times X_{\tau 2}$ in (2.10a) and also for $[X^2_{\tau 2}, -X^1_{\tau 1}]$ in (2.10b) to get (2.12) for $m = 2, 3$.

Differentiating (2.12) and (2.13), we obtain

$$(2.14)\qquad (nX^i_\sigma)_\sigma = \frac{1}{2n}\frac{\partial n}{\partial x^i} \quad \text{or} \quad (nX_\sigma)_\sigma = \frac{\operatorname{grad} n}{2n}\ .$$

These are the differential equations for the geodesics determined by the metric

$$(2.15)\qquad ds = n^{\frac{1}{2}}|dx|\ ;$$

see [4].

We summarize the above in a theorem.

Theorem 2.5. If X defines a ray coordinate system (Definition 2.1), then the curves $\{x | x = X(\sigma, \tau)\ ,\ \sigma \geq 0,\ \tau = \text{fixed},\ \tau \in D'\}$ are optical paths (geodesics in the Riemannian metric given by (2.15), where $n^{\frac{1}{2}} = |X_\sigma|^{-1}$).

This result gives the solution to part of the third problem posed in the first paragraph of Section 2.1. Further on in this Section, and in Sections 5 and 6, we complete the solution to this problem by showing that fields of geodesics of the Riemannian metric given by (2.15) define ray coordinate systems.

Integrating (2.14) with respect to σ and using the orthogonality relations (2.7), we conclude that if

$$(2.16)\qquad \begin{aligned} &X(\sigma, \tau) = \nu(\tau)\sigma + X^0(\tau) + \mathcal{O}(\sigma^{-\epsilon}), \\ &X_\sigma(\sigma, \tau) = \nu(\tau) + \mathcal{O}(\sigma^{-1-\epsilon}), \quad \text{and} \quad X_{\sigma\sigma}(\sigma, \tau) = \mathcal{O}(\sigma^{-2-\epsilon}) \\ &\qquad (\sigma \to \infty) \end{aligned}$$

for some $\epsilon > 0$, then

$$nX_\sigma = \nu(\tau) - \int_\sigma^\infty (\nabla n/2n)dt$$

or

$$(2.17)\qquad X_\sigma - \nu(\tau) = (-1 + n^{-1})\nu(\tau) - n^{-1}\int_\sigma^\infty \frac{\nabla n}{2n}\, dt .$$

Here $\nu(\tau)$ is the exterior unit normal to $\hat{B} = \{x \mid x = X^0(\tau),\ \tau \in D\}$; we assume that $\hat{B}$ is locally convex.

If the conditions (2.16) hold, then integration of (2.17) from σ to ∞ yields the result

$$(2.18)\qquad X(\sigma, \tau) = \nu(\tau)\sigma + X^0(\tau) - \nu(\tau)\int_\sigma^\infty [-1 + n^{-1}]dt + \int_\sigma^\infty n^{-1}\int_t^\infty \frac{\nabla n}{2n}\, ds\, dt$$

which is an integral equation for the rays. We have thus proved

Lemma 2.6. *If X determines an orthogonal ray coordinate system (ray system) and the conditions (2.16) are satisfied for some $\epsilon > 0$, then the integral equation (2.18) is satisfied by X, where n is defined to be $|X_\sigma|^{-2}$.*

In the remainder of this Section we assume that it is the index of refraction $n^{\frac{1}{2}}(x)$ which is given and that X is a C^2-solution of (2.18) such that

$$(2.19)\qquad \begin{aligned} &X = \nu(\tau)\sigma + X^0(\tau) + \mathcal{O}(\sigma^{-1}), \\ &X_\sigma = \nu(\tau) + \mathcal{O}(\sigma^{-2}), \qquad\qquad (\sigma \to \infty) . \\ &X_{\tau i} = \nu_{\tau i}(\tau)\sigma + X^0_{\tau i}(\tau) + \mathcal{O}(\sigma^{-1}) \end{aligned}$$

We make the following assumptions on n:

Hypotheses

$$(2.20)\qquad n \in C^{2N+6}(\mathbb{R}^m) \qquad (m = 2, \text{ or } 3) ;$$

and there exist positive constants n_0 and C such that on $\mathbb{R}^m$

$$(2.21)\qquad n(x) \geq n_0 ,$$

(2.22) $$\sup[|x|^2 \, |n(x)-1|] < C ,$$

(2.23) $$\sup[|x|^{2+|p|} |D^p n(x)|] < C , \quad (1 \le |p| \le 2N+6) ,$$

where $p = (p_1, p_2, \ldots, p_m)$, $p = \Sigma p_i$, *the* p_i *are nonnegative integers, and* N *is a positive integer.*

In this Section and in Section 4, 5, and 6 we only use the smoothness and asymptotic behavior of $D^p n$ for $|p| = 0$, 1, and 2. The full power of the Hypotheses is used in Section 7 and 8, as well as in Chapter 3. The existence of X satisfying (2.18) is proved in Section 4, and the above asymptotic behavior (2.19) of X_σ and $X_{\tau i}$ is established in Section 6, Theorem 6.1. Under the hypotheses (2.19) we next show that the orthogonality relations are satisfied and that $|X_\sigma| = n^{\frac{1}{2}}$.

First we prove that $|X_\sigma| = n^{\frac{1}{2}}$. It is easy to obtain the differential equations (2.14) from the integral equation (2.18). Thus

(2.24) $$X_{\sigma\sigma} = \frac{-n_\sigma}{n} X_\sigma + \frac{\nabla n}{2n^2} ,$$

from which it follows that,

$$(X_\sigma \cdot X_\sigma)_\sigma = 2X_\sigma \cdot \left[-\frac{n_\sigma}{n} X_\sigma + \frac{\nabla n}{2n^2} \right] .$$

For brevity we now write $n(\sigma)$ for $n(X(\sigma,\tau))$. Integrating both sides of the above expression from s to σ, we conclude that

(2.25) $$|X_\sigma(\sigma,\tau)|^2 = |X_\sigma(s,\tau)|^2 \left[\frac{n(s)}{n(\sigma)}\right]^2 + n^{-2}(\sigma) \int_s^\sigma n_\sigma(t)dt$$

$$= n^{-2}(\sigma)[n^2(s)\,|X_\sigma(s,\tau)|^2 - n(s)] + n^{-1}(\sigma) .$$

We now let $s \to \infty$, and we observe that by (2.22) and (2.19) both $n(s)$ and $|X_\sigma(s,\tau)|$ approach 1. Thus (2.25) becomes

(2.26) $$|X_\sigma(\sigma,\tau)|^2 = n^{-1}(X(\sigma,\tau)) ,$$

which is what we wished to prove.

The proof that $X_\sigma \cdot X_{\tau i} = 0$ is also easy. Using (2.24) and (2.26), we obtain

(2.27) $$(X_\sigma \cdot X_{\tau i})_\sigma = \left(-\frac{n_\sigma}{n} X_\sigma + \frac{\nabla n}{2n^2}\right) \cdot X'_{\tau i} - \frac{n_{\tau i}}{2n^2} .$$

We note that $n_{\tau i} = \nabla n \cdot X_{\tau i}$. Thus (2.27) reduces to

(2.28) $$(X_\sigma \cdot X_{\tau i})_\sigma = -\frac{n_\sigma}{n}(X_\sigma \cdot X_{\tau i}) \quad (i = 1, m-1) .$$

Solving (2.28), we obtain the relation

(2.29) $$X_\sigma(\sigma,\tau) \cdot X_{\tau i}(\sigma,\tau) = X_\sigma(s,\tau) \cdot X_{\tau i}(s,\tau) \frac{n(s)}{n(\sigma)} .$$

The hypotheses (2.22) on $n(x)$ and the asymptotic formulas (2.19) imply that as $s \to \infty$

$$n(s) \to 1 \quad \text{and} \quad X_\sigma(s, \tau) \cdot X_{\tau i}(s, \tau) \to 0 .$$

Therefore, (2.29) implies that

$$X_\sigma(\sigma, \tau) \cdot X_{\tau i}(\sigma, \tau) = 0 \quad \text{for all} \quad (\sigma, \tau) \in \mathcal{S} \quad (i = 1, m-1) . \tag{2.30}$$

We summarize these results in

Theorem 2.7. *If* $n \in C^2(\mathbb{R}^m)$ *and satisfies the hypotheses* (2.21)-(2.23) *for* $|p| = 0, 1,$ *and* 2 *and* $\mathcal{S}_M$ *is a subregion of* $\mathcal{S}$ *of the form* $\{(\sigma, \tau) | \sigma \geq M, \tau \in D'\}$ *such that* X *is a* C^2*-solution of* (2.14) *with the properties* (2.19) *on* $\mathcal{S}_M$, *then*

$$X_\sigma \cdot X_\sigma = n^{-1}(X) \quad \text{and} \quad X_\sigma \cdot X_{\tau i} = 0 \quad (i = 1, 2) .$$

It follows from (2.19) that $J > 0$ for $\sigma \geq M$ if M is sufficiently large.

In Section 5 we show that

$$(\sigma, \tau) \to X(\sigma, \tau)$$

is a one-to-one mapping of $\mathcal{S}^*_{\sigma_3}$ onto

$$\mathcal{R}_3 = \{x | x = X(\sigma, \tau), (\sigma, \tau) \in \mathcal{S}^*_{\sigma_3}\}$$

for some sufficiently large σ_3 ; see Theorem 5.1 and Corollary 5.2. Thus the C^2-solution of (2.18) that satisfies (2.19) yields a *ray system* on $\overline{T} = \mathcal{R}_3$, and

$$\Sigma_\sigma = \{x | x = X(\sigma, \tau), \tau \in D''\}$$

is a locally convex patch relative to $n^{\frac{1}{2}}$ for every $\sigma \geq \sigma_3$.

3. An Existence Theorem

Henceforth, except where it is stated otherwise, we assume that an index of refraction $n^{\frac{1}{2}}$ with the properties (2.20)-(2.23) has been given. Our goal is to demonstrate existence, uniqueness, smoothness, and asymptotic properties of a solution of the integral equation (2.18) and then to show that this solution defines a 1 - 1 mapping from $\mathcal{S}^*_M$, for some $M > 0$, onto an infinitely long closed tube $\overline{T}$ in $\mathbb{R}^m$. If we do this, then by Theorem 2.7 we may conclude that the solution X of (2.18) determines a *ray system* corresponding to the given $n^{\frac{1}{2}}$ on $\overline{T}$.

We begin by proving a general existence uniqueness theorem for nonlinear integral equations of a certain class (chosen with (2.18) in mind). We consider integral equations of the form

$$y(\sigma, \tau) = G(y; \sigma, \tau)$$

$$(3.1) \qquad = \int_\sigma^\infty K_1(u(s, \tau); s, \tau)\, ds + \int_\sigma^\infty K_2(u(s, \tau); s, \tau)\left(\int_s^\infty K_3(u(t, \tau); t, \tau)\, dt\right) ds \ ,$$

where $u(s, \tau) = y(s, \tau) + F(s, \tau)$. We assume that F is continuous on $\mathcal{S} = [0, \infty) \times \mathcal{T}$, where $\mathcal{T}$ is some compact domain in τ-space. We assume

$$(3.2) \qquad |G(y; \sigma, \tau)| \le \int_\sigma^\infty k_1(|u(s,\tau)|)ds + \int_\sigma^\infty\left(\int_s^\infty k_2(|u(t, \tau)|)dt\right) ds \ ,$$

where $k_j(t)$ $(j = 1, 2)$ are positive, strictly decreasing functions of t, for all $t > 0$. We shall also impose some integrability conditions on the k_j see (3.8) and (3.12) below. Further, we restrict our attention to equations of the form (3.1) in which the vector F satisfies the inequality

$$(3.3) \qquad |F(\sigma, \tau)| \ge 2^{-\frac{1}{2}}[g(\sigma) + h(\tau)]$$

for some g and h with the properties

$$(3.4) \qquad \begin{array}{ll} \text{(i)} & g(0) = 0, \ g'(\sigma) \ge g_0 > 0 \quad \text{for} \quad \sigma \ge 0 \\ \text{(ii)} & h(\tau) \ge h_0 > 0 \quad \text{for} \quad \tau \in \mathcal{T} \ . \end{array}$$

We shall prove that, in an appropriate space S, sufficiently high iterates of G are contraction maps from S into S for all $\sigma \ge 0$. Hence we shall conclude that (3.1) has a unique solution $y \in S$.

By (3.2), the monotonicity of the k_j, and (3.3)

$$(3.5) \qquad |G(y; \sigma, \tau)| \le \int_\sigma^\infty k_1\left(\frac{g(s) + h(\tau)}{\sqrt{2}} - |y(s, \tau)|\right) ds + \int_\sigma^\infty \left[\int_s^\infty k_2\left(\frac{g(t) + h(\tau)}{\sqrt{2}} - |y(t, \tau)|\right) dt\right] ds \ .$$

Let $\delta \in (0, 2^{-\frac{1}{2}}]$, $v(\sigma) = \delta[g(\sigma) + h_0]$, and

$$(3.6) \qquad \Lambda(\sigma, \delta) = \int_\sigma^\infty k_1(v(s))ds + \int_\sigma^\infty\left[\int_s^\infty k_2(v(t))dt\right] ds \qquad (\sigma \ge 0) \ .$$

A calculation shows that since $g'(\sigma) \ge g_0 > 0$,

$$(3.7) \qquad 0 \le \Lambda(\sigma, \delta) \le M(\delta) \stackrel{d}{=} \frac{1}{\delta g_0}\left\{\int_{\delta h_0}^\infty k_1(s)ds + \frac{1}{\delta g_0}\int_{\delta h_0}^\infty \int_s^\infty k_2(t)dtds\right\} \ .$$

We assume that

$$(3.8) \qquad M(\delta) < \infty \quad \text{for all} \quad \delta \in (0, 2^{-\frac{1}{2}}] \ .$$

Let S be the space of all continuous functions

$$y\colon \mathbb{S} \to \mathbb{R}^m$$

which satisfy the inequality

$$(3.9) \qquad |y(\sigma, \tau)| \le \Lambda(\sigma, \delta) \quad (\sigma \ge 0) ,$$

uniformly in τ . By (3.4), (3.7), and (3.9)

$$2^{-\frac{1}{2}}[g(\sigma) + h(\tau)] - |y(\sigma, \tau)| \ge 2^{-\frac{1}{2}}[g(\sigma) + h_0] - M .$$

Therefore, if we can show that there exists a $\delta \in (0, 2^{-\frac{1}{2}})$ such that

$$(3.10) \qquad 1 - \frac{2^{\frac{1}{2}}}{h_0} M(\delta) = \delta 2^{\frac{1}{2}} ,$$

then it follows easily from (3.5) that

$$(3.11) \qquad |G(y; \sigma, \tau)| \le \Lambda(\sigma, \delta) ;$$

and we conclude that

$$G\colon S \to S$$

since $G(y; \sigma, \tau)$ is continuous on $\mathbb{S}$ if $y \in S$. We prove below (Lemma 3.1) that such a δ does exist, under suitable hypotheses on h_0 and the k_j $(j = 1, 2)$.

Let

$$\|y\| = \sup_{\mathbb{S}} |y(\sigma, \tau)| .$$

We now make our last assumptions on the "kernels" in (3.1).

Hypotheses.

(i) There exist positive constants H_1, H_2, and ϵ such that for all y_1 and y_2 in S

$$(3.12a) \qquad \begin{aligned} |G(y_2; \sigma, \tau) - G(y_1; \sigma, \tau)| \le {} & \int_\sigma^\infty H_1 v^{-1-\epsilon}(s) |y_2(s, \tau) - y_1(s, \tau)| ds \\ & + \int_\sigma^\infty \int_s^\infty H_2 v^{-2-\epsilon}(t) |y_2(t, \tau) - y_1(t, \tau)| dt ds , \end{aligned}$$

where $v(t) = \delta[g(t) + h(\tau)]$.

(ii) Given h_0 satisfying (3.4 -ii), then

$$(3.12b) \qquad 16 \cdot 2^{\frac{1}{2}} B h_0^{-1} < 1 ,$$

where

$$(3.13) \qquad B = g_0^{-1} \int_0^\infty k_1(s) ds + g_0^{-2} \int_0^\infty \left[\int_s^\infty k_2(t) dt \right] ds .$$

Using the hypotheses (3.12)-(3.13), we can easily show that for k sufficiently large, G^k is a contracting map from S into S ; hence it has a unique fixed point Y, and Y solves (3.1). Note that $\lim_{\sigma \to \infty} Y(\sigma, \tau) = 0$ uniformly in τ since

$\lim_{\sigma\to\infty} \Lambda(\sigma,\delta) = 0$.

It remains to prove the existence of a $\delta \in (0, 2^{-\frac{1}{2}})$ satisfying (3.10).

Lemma 3.1. *If the hypotheses* (3.2)-(3.4), (3.8), *and* (3.12b) *hold and the functions* k_j $(j = 1, 2)$ *are positive and strictly decreasing on* $[0, \infty)$, *then there exists a solution* δ *of* (3.10) *in* $(0, 2^{-\frac{1}{2}})$.

Proof. Set

$$A(\delta) = -\delta^3 2^{\frac{1}{2}} + \delta^2 - \frac{\sqrt{2}\,\delta}{h_0 g_0} \int_{\delta h_0}^{\infty} \left[k_1(s) + \frac{1}{g_0\delta} \int_s^{\infty} k_2(t)dt\right] ds \quad .$$

Then (3.10) becomes

$$A(\delta) = 0 \quad .$$

Note that $A(2^{-\frac{1}{2}}) < 0$ since the k_j are positive.

We shall show that there exists a δ in $(0, 2^{-\frac{1}{2}})$ such that $A(\delta) > 0$. This implies the existence of at least one solution of (3.10) in $(0, 2^{-\frac{1}{2}})$.
We begin by observing that

$$A(2^{-\frac{1}{2}} - \beta) = 2^{-\frac{1}{2}}\beta - 2\beta^2 + 2^{\frac{1}{2}}\beta^3 - \frac{2^{\frac{1}{2}}}{h_0} (2^{-\frac{1}{2}} - \beta)^2 M(2^{-\frac{1}{2}} - \beta) \ .$$

Let

$$\beta_0 = 2^{-5/2}[1 - (2^{9/2}Bh_0^{-1})^{\frac{1}{2}}] ,$$

where B is defined by (3.13). Note that β_0 is a zero of

$$A^*(\beta) = -2\beta^2 + 2^{-\frac{1}{2}}\beta - 2^{\frac{1}{2}}h_0^{-1}B$$

and

$$A(2^{-\frac{1}{2}} - \beta) = A^*(\beta) + P(\beta) ,$$

where $P(\beta) > 0$ if $\beta \in (0, 2^{-\frac{1}{2}})$. By our hypothesis (3.12b), $\beta_0 \in (0, 2^{-5/2})$. But

$$A(2^{-\frac{1}{2}} - \beta_0) = P(\beta_0) > 0 \quad .$$

Thus $A(\delta)$ has a zero between $2^{-\frac{1}{2}} - \beta_0$ and $2^{-\frac{1}{2}}$. This completes the proof of the lemma.

We summarize the main result of this Section as a theorem.

Theorem 3.2. *If the hypotheses* (3.2)-(3.4), (3.8) *and* (3.12) *hold and the functions* k_j *are positive and strictly decreasing on* $[0, \infty)$, *then the integral equation* (3.1) *has a unique continuous solution* Y *such that* $Y \in S$.

4. Solution of the Ray Equations

The system of ray equations (2.18) takes the form (3.1) with the obvious identifications

$$(4.1) \qquad K_1 = \nu(\tau)\,\frac{n(x)-1}{n(x)}\,, \quad K_2 = \frac{1}{n(x)}\,, \text{ and } K_3 = \frac{\operatorname{grad} n(x)}{2n(x)}$$

$$(x = X(\sigma,\tau))\,.$$

For the moment we choose X^0 to be any function such that

$$\hat{B} = \{x \mid x = X^0(\tau);\ \tau \in D\}$$

describes a smooth, bounded $(m-1)$-manifold in $\mathbb{R}^m$. Thus there exists an $h_0 > 0$ such that

$$h(\tau) \stackrel{d}{=} |X^0(\tau)| \geq h_0 \quad \text{for} \quad \tau \in D\,.$$

We choose $g(\sigma) = \sigma$. Then $g_0 = 1$, and the hypotheses (3.3) and (3.4) are satisfied with

$$(4.2) \qquad F(\sigma,\tau) = \nu\sigma + X^0(\tau)\,.$$

By our basic hypotheses (2.21) and (2.22), there exists a $C > 0$ such that

$$(4.3a) \qquad |K_1| \leq \frac{C}{n_0|x|^2} \stackrel{d}{=} k_1(|x|)\ ;$$

and if we set

$$(4.3b) \qquad k_2(|x|) \stackrel{d}{=} \frac{C}{2n_0^2|x|^3}\,,$$ then by hypotheses (2.21) and (2.23)

$$|K_3| \leq n_0 k_2(|x|)\,.$$

These bounds k_i satisfy the hypotheses of monotonicity in Theorem 3.2, and the hypotheses (3.2) and (3.8) are obviously fulfilled. Moreover, the hypothesis (3.12b) is satisfied if X^0 is chosen so that h_0 is large enough compared to both C/n_0 and C/n_0^2. If n is sufficiently close to 1 and the first derivatives of n are close to 0, then h_0 need not be very large. The smaller $\sup_{\mathbb{R}^m}[\,|1-n|,\ |\operatorname{grad} n|\,]$ is, the smaller we may choose h_0.

To complete our application of Theorem 3.2 to the system of integral equations (2.18), we must show that the hypothesis (3.12a) is satisfied. Our approach to this is straightforward, but there are several delicate points in the analysis.

We adopt the notation

$$(4.4) \qquad \Delta f(\sigma) = f(u_2(\sigma)) - f(u_1(\sigma))$$

for any function f. Then

$$G(y_2,\sigma) - G(y_1,\sigma) = -\nu\int_\sigma^\infty \Delta n^{-1}(t)dt + \tfrac{1}{2}\int_\sigma^\infty n^{-1}(u_2(t))\int_t^\infty \Delta[\nabla n(s)]n^{-1}(u_2(s))ds\,dt$$

$$+ \tfrac{1}{2}\int_\sigma^\infty n^{-1}(u_2(t))\int_t^\infty \nabla n(u_1(s))\Delta n^{-1}(s)ds\,dt$$

$$(4.5) \qquad + \tfrac{1}{2}\int_\sigma^\infty \Delta n^{-1}(t)\int_t^\infty \frac{\nabla n(u_1(s))}{n(u_1(s))}\,ds\,dt \ ,$$

$$\overset{d}{=} (I) + (II) + (III) + (IV) \ .$$

Now by (3.3),

$$|u(\sigma, \tau)| \overset{d}{=} |F(\sigma, \tau) + y(\sigma, \tau)|$$

$$\geq \left(\frac{\sigma + h(\tau)}{\sqrt{2}}\right)\left(1 - \frac{\sqrt{2}\ |y(\sigma, \tau)|}{h(\tau)}\right) \ .$$

But $y \in S$ implies that

$$(4.6) \qquad |y(\sigma, \tau)| \leq \Lambda(\sigma, \delta) \leq M(\delta).$$

Thus

$$|u_i(\sigma, \tau)| \geq \frac{\sigma + h(\tau)}{\sqrt{2}}\left(1 - \frac{\sqrt{2}\ M(\delta)}{h_0}\right)$$

$$(4.7) \qquad\qquad (i = 1,\ 2)$$

$$\geq (\sigma + h_0)\delta \ ,$$

by the definition of δ as a solution of (3.10) on $(0,\ 2^{-\frac{1}{2}})$, which exists by Lemma 3.1. In (4.7) the u_i are not components of u, but rather the points corresponding to the y_i in (3.12a).

Next we observe that since n is assumed to be smooth,

$$|\Delta n| \leq |\nabla n(u_1 + \theta(u_2 - u_1))|\ |u_2 - u_1| \ ,$$

$$(4.8) \qquad \leq 2n_0^2 k_2(|u_1 + \theta(u_2 - u_1)|)\ |u_2 - u_1|$$

by the hypotheses (2.20), (2.23) and by (4.3), where $\theta \in (0, 1)$. But by (3.3), (3.4), (3.7), (3.9), and (3.10),

$$|u_1 + \theta(u_2 - u_1)| = |F(\sigma, \tau) + (1 - \theta)y_1 + \theta y_2| \ ,$$

$$\geq \frac{\sigma + h(\tau)}{\sqrt{2}} - |(1 - \theta)y_1 + \theta y_2| \ ,$$

$$(4.9) \qquad \geq \frac{\sigma + h(\tau)}{\sqrt{2}} - M(\delta) \ ,$$

$$\geq \frac{\sigma + h(\tau)}{\sqrt{2}}\left(1 - \frac{\sqrt{2}\ M(\delta)}{h_0}\right),$$

$$= [\sigma + h(\tau)]\delta \overset{d}{=} v(\sigma, \tau) \ .$$

Since k_2 is a strictly decreasing function, we thus conclude that

$$(4.10) \qquad |\Delta n| \leqq 2n_0^2 k_2(v)|y_2 - y_1| .$$

We can now estimate the terms (I), (IV), and (III) in (4.5). By (4.7)-(4.10) we conclude that

$$(4.11) \qquad \begin{aligned} |I| &\le \int_\sigma^\infty \frac{\Delta n(t)}{n_0^2}\, dt \\ &\le 2\int_\sigma^\infty k_2(v(t))|y_2(t) - y_1(t)|dt , \end{aligned}$$

and

$$(4.12) \qquad |IV| \le 2n_0\int_\sigma^\infty k_2(v(t))|y_2(t) - y_1(t)|\int_t^\infty k_2(v(s))ds\, dt ,$$

where $v = [\sigma + h(\tau)]\delta$. Also

$$(4.13) \qquad |III| \le 2n_0 \int_\sigma^\infty \int_t^\infty k_2^2(v(s))|y_2(s) - y_1(s)|ds\, dt .$$

It remains to estimate the term (II) in (4.5). We begin by observing that

$$\begin{aligned} |\Delta(\nabla n(\sigma))| &\overset{d}{=} |\nabla n(u_2) - \nabla n(u_1)| \\ &\le \Sigma_1^m |\Delta(n_i(\sigma))| , \end{aligned}$$

where n_i is the derivative of n with respect to x^i . But with an appropriate choice of C in (4.3), we conclude from our hypotheses (2.20)-(2.23) on n that

$$(4.15) \qquad \begin{aligned} |\Delta(n_i(\sigma))| &\le |\nabla n_i(u_1 + \theta(u_2 - u_1))|\ |u_2 - u_1| \\ &\le k_3(|u_1 + \theta_i(u_2 - u_1)|)|u_2 - u_1| , \end{aligned}$$

where $k_3(u) = Cu^{-4}$. Thus, by (4.15) and (4.7)-(4.9),

$$(4.16) \qquad |II| \le \frac{m}{2n_0^2} \int_\sigma^\infty \int_s^\infty k_2(v(t))|y_2(t) - y_1(t)|dt\, ds .$$

It follows immediately from (4.11)-(4.13) and (4.16) that the hypothesis (3.12a) is satisfied with $\epsilon = 1$ and appropriate choices of H_1 and H_2 . The conclusion we draw from this and the previous results of this Section is the following corollary to Theorem 3.2.

<u>Proposition</u> 4.1. <u>Under the hypotheses of Theorem</u> 2.7 <u>on</u> n , <u>if</u> $\hat{B}$ <u>is a smooth bounded</u> $(m-1)$-<u>dimensional manifold in</u> $\mathbb{R}^m$ <u>and</u>

$$h_0 \overset{d}{=} \inf|X^0(\tau)| \quad (\tau \in D)$$

<u>is sufficiently large, then the system of ray equations</u> (2.18) <u>has a unique continuous solution</u> X <u>such that</u> $(X - F) \in S$, <u>where</u> S <u>is the space defined by the sentence containing</u> (3.9).

5. Existence of Ray Fields on Unbounded Domains

Let D' be a proper subregion of D, where as above D is a compact subset of $\{\tau | \tau \in [\tau_1, \tau_2)\}$. For each $t \geq 0$ we define

$$\mathcal{S}_t = \{(\sigma,\tau) \,|\, (\sigma,\tau) \in [t,\infty) \times D'\} \quad . \tag{5.1}$$

In this Section we first establish the existence of a half cylinder (strip) $\mathcal{S}^*_{\sigma_3} = \{(\sigma,\tau) \,|\, (\sigma,\tau) \in [\sigma_3,\infty) \times D''\}$ that is mapped in a one-to-one way onto the closure of an unbounded, connected, proper subset $\mathcal{R}_3$ of $\mathbb{R}^m$ by the solution X of (2.18) determined in Proposition 4.1. The region $\mathcal{R}_3$ is the closure of the infinitely long tubular region T bounded by the surface (curves)

$$\partial T - S = \{x | x \in \mathbb{R}^m , \ x = X(\sigma,\tau) \ , \ (\sigma,\tau) \in [\sigma_3,\infty) \times \partial D''\}$$

and the surface (curve)

$$S = \{x | x \in \mathbb{R}^m , \ x = X(\sigma_3,\tau) \ , \ \tau \in D''\}$$

for some sufficiently large σ_3, where D'' is a closed proper subset of D' ; see Fig. 8. We shall further specify D' and D'' below.

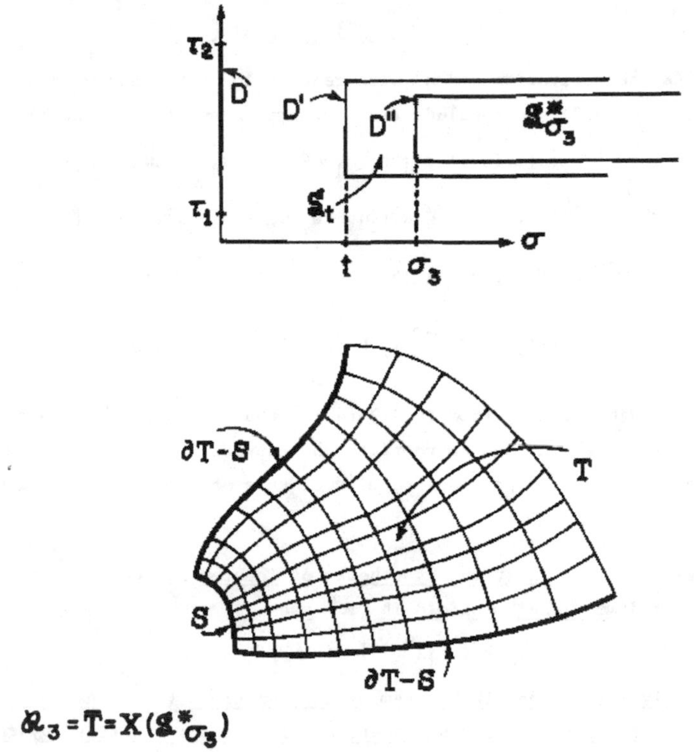

Figure 8

Henceforth we assume that the surface (curve)

$$\hat{B}_\sigma = \{x \mid x \in \mathbb{R}^m , x = F(\sigma,\tau) , \tau \in D\} , \tag{5.2}$$

where

$$F(\sigma,\tau) = \sigma\nu(\tau) + X^0(\tau) , \tag{5.3}$$

is locally convex and smooth for each $\sigma \geq 0$ and that D is simply connected. Here $\nu(\tau)$ is the unit normal emanating from the convex side of $\hat{B} = \hat{B}_0$. We assume also that the set of all straight lines

$$\hat{\mathfrak{F}} = [\hat{C}_\tau \mid \hat{C}_\tau = \{x \mid x = F(\sigma,\tau) , \sigma \geq 0\} , \tau \in D]$$

emanating normally from B forms a field on the sectorial region $\hat{T}$ bounded by $\hat{B}$ and the (m-1)-manifold

$$\partial\hat{T} - \hat{B} = \{x \mid x = F(\sigma,\tau) , (\sigma,\tau) \in [0,\infty) \times \partial D\} .$$

In the sequel we are particularly interested in the case where $\hat{B}$ is a patch of the boundary $\tilde{B} = \{x \mid x = \tilde{X}^o(\tau') , \tau' \in [\tau_1,\tau_2)\}$ of a convex body. ($\tilde{B}$ is a closed, convex, smooth embedding of an (m-1)-sphere in $\mathbb{R}^m$.) In such a case each set $\hat{B}$ is a locally convex patch on $\tilde{B}$, and there is a transformation $\tau' = \tau'(\tau)$ such that $\tilde{X}^o(\tau'(\tau)) = X^o(\tau)$ for all $\tau \in D$. We assume later that the curves $\{x \mid x = X^o(\tau) , \tau^i = \text{const.}\}$ on $\hat{B}$ are curves of constant principle curvature.

That there exists a 1 - 1 correspondence between $\mathcal{S}^*_{\sigma_3}$ and $\mathcal{R}_3$ under X implies that the set of all rays

$$\mathfrak{F} = [C_\tau \mid C_\tau = \{x \mid x = X(\sigma,\tau) , \sigma \geq \sigma_3\} , \tau \in D'']$$

forms a field $\mathfrak{F}$ on $\mathcal{R}_3$. A corollary of this result (see Corollary 5.3 proved below) is that if $\tilde{B}$ is a closed, convex, smooth embedding of an (m-1)-sphere in $\mathbb{R}^m$, then there exists a neighborhood $\mathfrak{n}$ of ∞ on which the set of all rays

$$\tilde{\mathfrak{F}} = [C_{\tau'} \mid C_{\tau'} = \{x = \tilde{X}(\sigma,\tau') , \sigma \geq \sigma_3\} , \tau' \in [\tau_1,\tau_2)]$$

forms a field; see Fig. 9. Here $\tilde{X}(\sigma,\tau')$ is the solution of the integral equation

$$\tilde{X} = \tilde{F}(\sigma,\tau') + \tilde{\mathfrak{F}}(\tilde{X}(\cdot,\tau') ; \sigma,\tau')$$

where

$$\tilde{F}(\sigma,\tau') = \tilde{\nu}(\tau')\sigma + \tilde{X}^0(\tau') ,$$

$$\tilde{\mathfrak{F}}(\tilde{X}(\cdot,\tau');\sigma,\tau') = -\tilde{\nu}(\tau')\int_\sigma^\infty [-1+n^{-1}(\tilde{X}(s,\tau'))]ds + \int_\sigma^\infty n^{-1}(\tilde{X}(s,\tau'))\int_s^\infty \frac{\nabla n(\tilde{X}(t,\tau'))}{2n(\tilde{X}(t,\tau'))} dtds ,$$

and $\tilde{\nu}(\tau')$ is the outward unit normal to $\tilde{B}$. By definition $\tilde{F}(\sigma,\tau'(\tau)) = F(\sigma,\tau)$, and $\tilde{\mathfrak{F}}(\tilde{X}(\cdot,\tau'(\tau));\sigma,\tau'(\tau)) = \mathfrak{F}(X(\cdot,\tau);\sigma,\tau)$ for all $\tau \in D$, where $\mathfrak{F}(X(\cdot,\tau);\sigma,\tau)$ comprises the integral terms in (2.18). It is obvious that $\tilde{X}(\sigma,\tau'(\tau)) \equiv X(\sigma,\tau)$ for all $(\sigma,\tau) \in [\sigma_3,\infty) \times D$, where $X(\sigma,\tau)$ is the unique solution of

$$X(\sigma,\tau) = F(\sigma,\tau) + \mathfrak{F}(X(\cdot,\tau);\sigma,\tau) \tag{5.4}$$

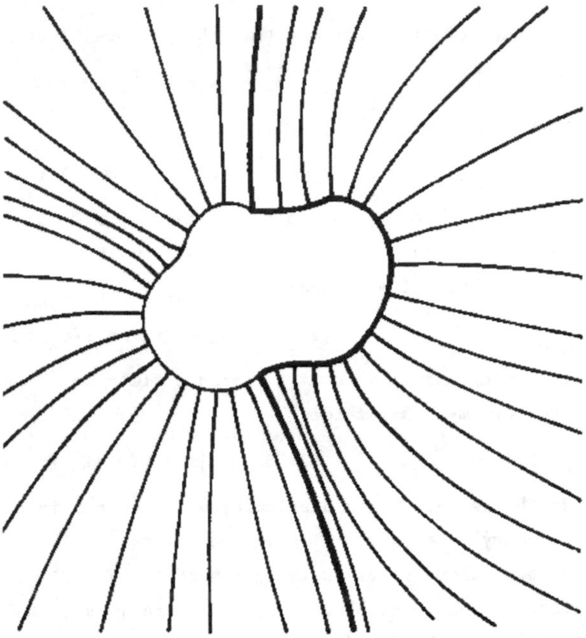

Figure 9

for all $\sigma \geq \sigma_3$, $\tau \in D$. Consequently, $\tilde{\mathfrak{J}} \supseteq \mathfrak{J}$.

Again we assume that the index of refraction $n^{\frac{1}{2}}$ is given and satisfies the hypotheses of Theorem 2.7.

Because of our assumption that $\hat{B}$ is smooth and convex, the map

$$F : \hat{\mathcal{S}} \to \hat{T} \cup \partial\hat{T}$$

is $1-1$ and onto, where

$$\hat{\mathcal{S}} = \{(\sigma,\tau) \mid \sigma \geq 0 , \tau \in D\} .$$

We let F^{-1} be the inverse of F. Since $\mathfrak{J} \to 0$ as $\sigma \to \infty$ because of (3.2), (3.3), (3.5), (3.7), (3.8), and (3.11), we expect that the map $(\sigma,\tau) \to (F+\mathfrak{J})(\sigma,\tau)$ will be $1-1$ from a subregion $\mathcal{S}_2$ of $\hat{\mathcal{S}}$ onto an unbounded domain $\mathcal{R}_2 \subset \mathbb{R}^m$ such that

$$\inf_{(\sigma,\tau) \in \mathcal{S}_2} \sigma$$

is sufficiently large. We establish the existence of $\mathcal{S}_2$ and $\mathcal{R}_2$ as follows.

Define

$$\rho = \sup |\mathfrak{J}(X(\cdot,\tau);\sigma,\tau)| \text{ , where } (\sigma,\tau) \in \mathcal{S}_1 , \tag{5.5}$$
$$\mathcal{S}_1 = \{(\sigma,\tau) \mid (\sigma,\tau) \in \hat{\mathcal{S}} , \sigma \geq \sigma_1\} ,$$

where σ_1 is chosen below and X is the previously found solution of the ray equations (2.18). Recall that, by virtue of (3.11), $|\mathfrak{J}| \leq \Lambda(\sigma,\delta)$, where $\Lambda(\sigma,\delta)$

is defined by (3.6). Having selected X^0 in (5.2) and (5.3), choose a subdomain $\mathcal{R}_1$ of $\hat{T}$ so that

$$x - \rho u \in \bar{\hat{T}} \quad \text{if} \quad x \in \mathcal{R}_1 ,$$

for every unit vector u. This can be accomplished since $\mathfrak{F} \to 0$ as $\sigma \to \infty$ (or as $h_0 \to \infty$). We define $\mathcal{R}_1$ to be the closure of the region bounded by $\hat{B}_{\sigma_1}$ and the surface (curves)

$$\{x | x \in \mathbb{R}^m , x = F(\sigma,\tau) , (\sigma,\tau) \in [\sigma_1,\infty) \times \partial D'\} .$$

We choose σ_1 to be a number so large that $\sigma_1 > \rho$ (ρ is defined in (5.5)). We next choose $D' = D'(\sigma_1)$ to be any subset of D such that

$$x - \rho u \in \bar{\hat{T}}$$

for every unit vector u, whenever

$$x \in \{x | x = F(\sigma,\tau) , (\sigma,\tau) \in [\sigma_1,\infty) \times \partial D'\} \cup \{x | x = F(\sigma,\tau) , \tau \in D'\} .$$

This condition will be satisfied if the distance of every point on $\partial\mathcal{R}_1$ from $\partial\hat{T}$ is at least ρ ; see Fig. 10.

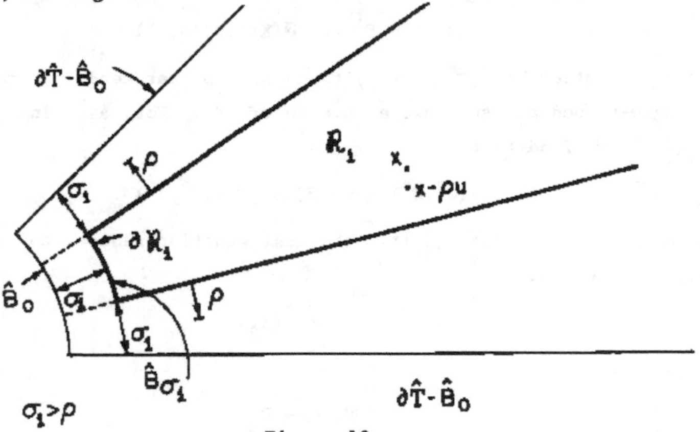

Figure 10

Therefore, if $x \in \mathcal{R}_1$, then $F^{-1}(x - ru)$ is well-defined for all unit vectors u and every $r \in [0,\rho]$. It follows that

$$(5.6) \qquad T_x(\sigma,\tau) = F^{-1}(x - \mathfrak{F}(X(\cdot,\tau);\sigma,\tau))$$

is well-defined for all $(\sigma,\tau) \in \mathcal{S}_{\sigma_1}$. Moreover if we define $\mathcal{R}_2$ to be the closed subregion of $\mathcal{R}_1$ bounded by the curve $\hat{B}_{\sigma_2}$, where $\sigma_2 = \sigma_1 + \rho$, and the surface (curves)

$$\{x | x \in \mathbb{R}^m , x = F(\sigma,\tau) , (\sigma,\tau) \in [\sigma_2,\infty) \times \partial D'\} ,$$

then $T_x : \mathcal{S}_{\sigma_1} \to \mathcal{S}_{\sigma_1}$ if $x \in \mathcal{R}_2$; see Fig. 11.

We know by Proposition 4.1 that, given $(\sigma,\tau) \in \mathcal{S}$, if h_0 is large enough, then

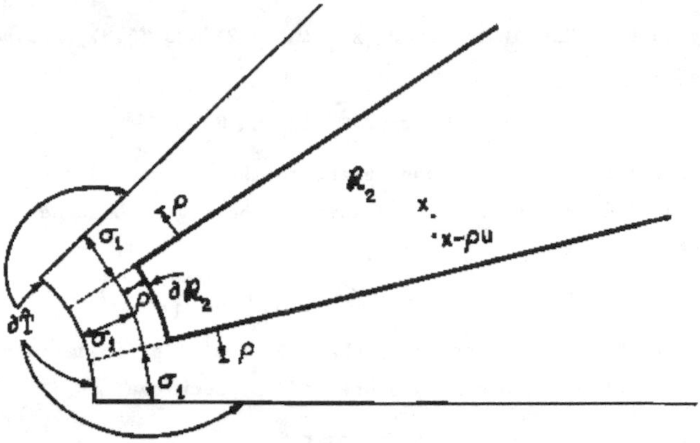

Figure 11

$\sigma_1 > \rho$

a unique $X(\sigma,\tau)$ can be found that satisfies (2.18) and the condition $(X - F) \in S$. Insofar as we know at this point, the inverse transformation X^{-1} from x to (σ,τ) may not be single-valued. But if $x \in \mathcal{R}_2$ and we can prove that the equations

$$(5.7) \qquad (\sigma,\tau) = F^{-1}(x - \mathfrak{F}(X(\cdot,\tau);\sigma,\tau))$$

have a unique solution $(\sigma^*,\tau^*) \in \mathcal{S}_{\sigma_1}$, then the map that sends x to (σ^*,τ^*) will be both single-valued and the unique inverse of X. For, applying F to both sides of (5.7), we find that

$$F(\sigma^*,\tau^*) = x - \mathfrak{F}(X(\cdot,\tau^*);\sigma^*,\tau^*) .$$

Thus, since $X(\sigma^*,\tau^*)$ solves (2.18), the last equation reduces to $x = X(\sigma^*,\tau^*)$. Therefore, if we define

$$(5.8) \qquad \mathcal{S}_2 = X^{-1}(\mathcal{R}_2) ,$$

then the mapping

$$(5.9) \qquad \mathcal{S}_2 \xrightarrow{X} \mathcal{R}_2$$

is $1 - 1$ and onto; see Fig. 12. We are thus motivated to prove the following theorem:

Theorem 5.1. *If* (1) $n \in C^2(\mathbb{R}^m)$ *and satisfies the hypotheses* (2.21)-(2.23) *for* $|p| = 0, 1,$ *and* 2, (2) h_0 *(defined in Proposition* 4.1) *is sufficiently large for Proposition* 4.1 *to hold, and* (3) σ_1 *is sufficiently large, then to each* $x \in \mathcal{R}_2$, *where* $\mathcal{R}_2$ *is as defined above, there corresponds a unique solution* $(\sigma^*,\tau^*) \in \mathcal{S}_{\sigma_1}$ *of* (5.7) *where* $\mathcal{S}_{\sigma_1}$ *is defined by* (5.1). *Further, the mapping* (5.9) *is a* $1 - 1$ *mapping from* $\mathcal{S}_2$ *onto* $\mathcal{R}_2$, *where* $\mathcal{S}_2$ *is defined by* (5.8).

Proof. In view of the preceding discussion we need only prove that the mapping $T : \mathcal{S}_{\sigma_1} \to \mathcal{S}_{\sigma_1}$ defined by the right-hand side of (5.7) is a contraction for each $x \in \mathcal{R}_2$. We use the metric

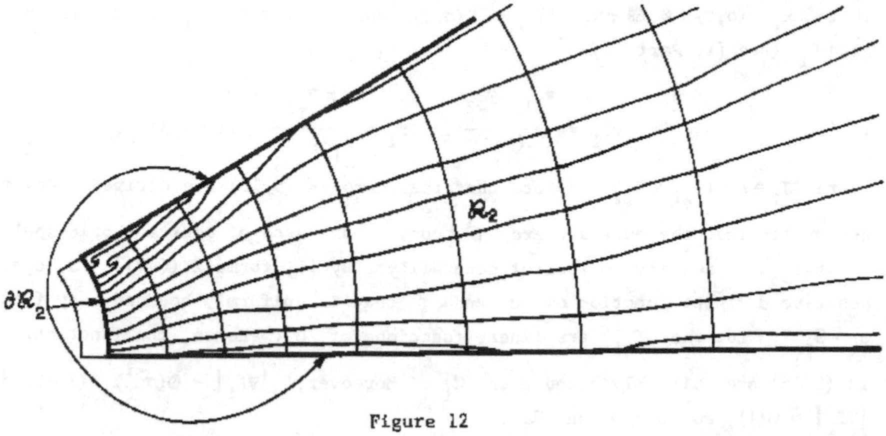

Figure 12

$$\rho((\tilde{\sigma},\tilde{\tau}),(\sigma,\tau)) \overset{d}{=} [|\tilde{\sigma}-\sigma|^2+\Sigma_1^{m-1}|\tilde{\tau}^i-\tau^i|^2]^{\frac{1}{2}} ,$$

and we write

$$(5.10)\qquad \begin{aligned} R(\sigma,\tau) &= \mathfrak{F}(X(\cdot,\tau);\sigma,\tau) , \\ T(\sigma,\tau) &= T_2 \circ T_1(\sigma,\tau;x) , \end{aligned}$$

where we have suppressed the dependence of T on x and where

$$(5.11)\qquad T_1(\sigma,\tau;x) = x - R(\sigma,\tau) \quad\text{and}\quad T_2(x) = F^{-1}(x) .$$

In the remainder of this proof we consider the three-dimensional case only $(m=3)$. The case $m=2$ is much simpler, and we leave it to the reader.

Let f_i be the ith component of F^{-1}, where $\sigma = f_1(x)$, $\tau^1 = f_2(x)$, and $\tau^2 = f_3(x)$. For $x \in \mathcal{R}_2$, we consider

$$(5.12)\qquad \begin{aligned} \Delta T &\overset{d}{=} |T(\tilde{\sigma},\tilde{\tau}) - T(\sigma,\tau)| \\ &= \{\Sigma_1^3 [f_i(x-R(\tilde{\sigma},\tilde{\tau})) - f_i(x-R(\sigma,\tau))]^2\}^{\frac{1}{2}} \\ &\overset{d}{=} [\Sigma_1^3 (\Delta f_i)^2]^{\frac{1}{2}} , \end{aligned}$$

where (σ,τ) and $(\tilde{\sigma},\tilde{\tau})$ lie in $\mathcal{S}_{\sigma_1}$ so that $|R(\sigma,\tau)|$ and $|R(\tilde{\sigma},\tilde{\tau})|$ are both bounded by ρ, and both $x - R(\tilde{\sigma},\tilde{\tau})$ and $x - R(\sigma,\tau)$ lie in $\mathcal{R}_1$. We show that if $x \in \mathcal{R}_2$, there exists an $\alpha \in (0,1)$ such that

$$(5.13)\qquad |\Delta T| \le \alpha\rho((\tilde{\sigma},\tilde{\tau}),(\sigma,\tau))$$

for all (σ,τ), $(\tilde{\sigma},\tilde{\tau})$ in $\mathcal{S}_{\sigma_1}$. By the smoothness of F^{-1} and the mean-value theorem, we know that there exist $\phi_i (0 < \phi_i < 1 ; i=1,2,3)$ such that

$$(5.14)\qquad |\Delta f_i| = |\operatorname{grad} f_i(x - R(\sigma,\tau) + \phi_i \Delta R)|\ |\Delta R| \quad (i=1,2,3) ,$$

where

$$\Delta R = R(\tilde{\sigma},\tilde{\tau}) - R(\sigma,\tau) .$$

(Note $x - R(\sigma,\tau) + \phi_i \Delta R \in \mathcal{R}_1$ if $x - R(\sigma,\tau)$ and $x - R(\tilde\sigma,\tilde\tau) \in \mathcal{R}_1$.) One can compute the grad f_i (see [1, Part II]):

$$\nabla f_1 = \frac{F_{\tau^1} \times F_{\tau^2}}{J_f(\sigma,\tau)} , \quad \nabla f_i = \frac{\nu \times F_{\tau^i}}{J_f(\sigma,\tau)} \quad (i = 2,3) , \tag{5.15}$$

where $J_f = \nu \cdot (F_{\tau^1} \times F_{\tau^2})$. Note that the formulas (5.15) are derived under the assumption that the curves where $\tau^i = \text{const.}$ are arcs of constant principal curvature on $\hat{B}$. This entails no loss of generality. By [1, formula (6.2)], $J_f(\sigma,\tau)$ is positive definite function of σ and τ that is uniformly bounded away from zero on $\mathcal{S}_{\sigma_1}$. But the F_{τ^i} are linear functions of σ . Hence, the functions $|\nabla f_i|$ in (5.15) are uniformly bounded on $\mathcal{R}_1$. Moreover, $|\nabla f_i| = \mathcal{O}(\sigma^{-1})$ $(i = 2,3)$ and $|\nabla f_1| = \mathcal{O}(1)$ as $\sigma \to \infty$ on $\mathcal{R}_1$.

Thus to show that T is a contraction, it is sufficient to show that

$$|\Delta R| \leq \beta \rho ((\tilde\sigma,\tilde\tau),(\sigma,\tau)) , \tag{5.16}$$

where we can make β as small as we please by choosing h_0 (or σ_1) sufficiently large. We only sketch the proof of (5.16). It is rather similar to the proof of Proposition 4.1. First, by adding and subtracting terms and using the lower bound n_0 on $n(x)$ (see (2.21)), we easily show that

$$\begin{aligned} |\Delta R| \leq \frac{|\Delta\nu|}{n_0} \int_{\tilde\sigma}^{\infty} |n(X(t,\tilde\tau)) - 1|dt + n_0^{-1}\int_{\tilde\sigma}^{\infty} |n(X(t,\tilde\tau)) - n(X(t,\tau))|dt \\ + n_0^{-1}\left|\int_{\tilde\sigma}^{\sigma} |n(X(t,\tau)) - 1|dt\right| + n_0^{-1}\left|\int_{\tilde\sigma}^{\sigma}\int_{t}^{\infty} |\nabla n(X(s,\tilde\tau))|ds\, dt\right| \\ + n_0^{-3}\int_{\sigma}^{\infty} |n(X(t,\tilde\tau)) - n(X(t,\tau))| \int_{t}^{\infty} |\nabla n(X(s,\tilde\tau))|ds\, dt \\ + n_0^{-2}\int_{\sigma}^{\infty}\int_{t}^{\infty} |\nabla n(X(s,\tilde\tau)) - \nabla n(X(s,\tau))|ds\, dt \end{aligned} \tag{5.17}$$

We label the terms on the right-hand side of (5.17) I - VI, in the order they are written. It is straightforward to show that

$$|\Delta\nu|^2 \leq \sum_{i=1}^{3} \sum_{j=1}^{2} \left| \sup_{\tau\in D}\left(\frac{\partial\nu^i}{\partial\tau^j}\right)^2 \right| |\tilde\tau - \tau|^2 . \tag{5.18}$$

The suprema in (5.18) exist since the surface defined by X^0 is a bounded set in $\mathbb{R}^3$. Moreover by the hypotheses (2.20) - (2.23),

$$n_0^{-1}\int_{\tilde\sigma}^{\infty} |n(X(t,\tilde\tau)) - 1|dt \leq \int_{\tilde\sigma}^{\infty} k_1(|X(t,\tilde\tau)|)dt , \tag{5.19}$$

where k_1 is defined in (4.3a) at the beginning of Section 4. Since by (4.7)

$$|X(\sigma,\tau)| \geq [\sigma + h(\tau)]\delta , \tag{5.20}$$

the integral in (5.19) can be made as small as we like by choosing h_0 or σ_1 large.

Thus

(5.21) $$|I| \le \beta_I |\tilde{\tau} - \tau| ,$$

where β_I can be made as small as we desire by choosing h_0 or σ_1 sufficiently large. In what follows all constants β_i $(i = II,\ldots,VI)$ have the same property as β_I; namely, they can be made as small as we like by choosing h_0 or σ_1 large.

From the estimate (5.19) (with the upper limits of integration now being σ instead of ∞), the specific form of k_1, and the estimate (5.20), we can show that $|III| \le \beta_{III}|\tilde{\sigma} - \sigma|$.

Next, using the hypothesis (2.23) for $|p| \le 2$ in the form (4.3b) and (4.7), we can estimate grad n and obtain a bound $|IV| \le \beta_{IV}|\tilde{\sigma} - \sigma|$.

We consider the term II. First we observe, using the mean-value theorem and elementary inequalities, that

(5.22) $$\begin{aligned} \Delta n &\overset{d}{=} |n(X(t,\tilde{\tau})) - n(X(t,\tau))| \\ &\le |\nabla n(X(t,\tilde{\tau}) + \theta\,\Delta X)| \left[\sum_{i=1}^{3}\sum_{j=1}^{2}\left(\frac{\partial X^i}{\partial \tau^j}\right)^2\right] |\tilde{\tau} - \tau| \end{aligned}$$

for some $\theta \in (0,1)$, where $\partial X^i/\partial\tau^j$ is evaluated at $(t, \tau + \phi_i(\tilde{\tau} - \tau))$ and $\phi_i \in (0,1)$. By the same reasoning we used to derive (4.8), we obtain the following inequality from (5.22):

(5.23) $$|\Delta n| \le \text{Const.}\, k_2(u) \left[\sum_{j=1}^{3}\sum_{j=1}^{2}\left(\frac{\partial X^i}{\partial \tau^j}\right)^2\right]^{\frac{1}{2}} |\tilde{\tau} - \tau| ,$$

where $u = |X(t,\tilde{\tau}) + \theta\,\Delta X|$. We prove in Section 6 below (Theorem 6.1) that $\partial X^i/\partial\tau^j = \mathcal{O}(t)$ as $t \to \infty$, uniformly for $\tau \in D$. By (4.3b), $k_2(u) = \mathcal{O}(u^{-3})$ as $u \to \infty$. Thus if we can show that $u \ge (\text{Const.})t$, we easily obtain an inequality

(5.24) $$|II| \le \beta_{II}|\tilde{\tau} - \tau|$$

using (5.23).

We can bound u from below by a multiple of t if $|\tilde{\tau} - \tau| \le \delta_1$ for some appropriately small δ_1. To see this we compute, since $X(t,\tau) - F(t,\tau) = \mathcal{O}(t^{-1})$, that

$$|u|^2 = |\nu(\tilde{\tau}) + \theta\,[\nu(\tau) - \nu(\tilde{\tau})]|^2 t^2 + \mathcal{O}(t)$$

for some $\theta \in (0,1)$. But ν is a smooth function of τ by hypothesis. Thus, for some $\delta_1 > 0$, $|\nu(\tilde{\tau}) - \nu(\tau)| \le \frac{1}{2}$ if $|\tilde{\tau} - \tau| \le \delta_1$, uniformly for $\tau, \tilde{\tau} \in D$. Therefore by the triangle inequality, we find that

(5.25) $$|u|^2 \ge 3t^2/4 + \mathcal{O}(t) \qquad (|\tilde{\tau} - \tau| \le \delta_1) .$$

Hence (5.24) holds if $|\tilde{\tau} - \tau| \le \delta_1$.

For $|\tilde{\tau} - \tau| > \delta_1$, the integrand of the term II in (5.17) can be estimated directly:

$$|\Delta n| \le \frac{1}{\delta_1}\,[\,|n(X(t,\tilde{\tau})) - 1| + |n(X(t,\tau)) - 1|\,]\,|\tilde{\tau} - \tau| ,$$

from which (5.24) follows immediately by the hypothesis (2.22) and the estimate (5.20). Therefore (5.24) holds uniformly for all $\tau, \tilde{\tau} \in D$.

To estimate the term V in (5.17) we again use (i) hypothesis (2.23) with $|p| \leq 2$ to bound $|\text{grad } n|$ and (ii) our previous estimates of $|\Delta n|$ to find that: $|V| \leq \beta_V |\tilde{\tau} - \tau|$, uniformly for all allowed choices of τ and $\tilde{\tau}$, all $\tilde{\sigma} > 0$ and all sufficiently large σ .

Finally, the estimation of the term VI in (5.17) is conducted by an argument similar to the above, in which the estimate (4.15) is used. The result is: $|VI| \leq \beta_{VI} |\tilde{\tau} - \tau|$ uniformly for all allowed choices of τ and $\tilde{\tau}$, and all sufficiently large σ .

The above estimates taken together imply that

(5.26) $$|\Delta R| \leq [\beta_I + \beta_{II} + \beta_V + \beta_{VI}] |\tilde{\tau} - \tau| + [\beta_{III} + \beta_{IV}] |\tilde{\sigma} - \sigma| \ .$$

By Cauchy's inequality, (5.26) implies (5.16) (and hence (5.13)) where β , and hence α , can be made to lie as close to 0 as we please by choosing h_0 or σ_1 sufficiently large. By the reasoning given at the beginning of this Section this completes the proof of the theorem.

<u>Remark</u>. It is apparent from the proof of the Theorem 5.1 that there is a choice of hypotheses. Either h_0 may be chosen sufficiently large or σ_1 may be chosen sufficiently large and h_0 large enough to satisfy the conditions needed for Proposition 4.1 to hold.

We next prove that $\mathcal{S}_2 = X^{-1}(\mathcal{R}_2)$ contains a half-cylinder

$$\mathcal{S}^*_{\sigma_3} = \{(\sigma, \tau) | \sigma \geq \sigma_3 , \tau \in D''\}$$

for sufficiently large σ_3 and a properly chosen subset D'' of D' . To do this we choose $\sigma_3 = \sigma_1 + 2\rho$, where ρ is defined by (5.5), and we take D'' to be a subset of D' such that every point on $\partial\mathcal{R}_3$ is a distance ρ from $\partial\mathcal{R}_2$. (This means every point of $\partial\mathcal{R}_3 = X(\mathcal{S}^*_{\sigma_3})$ is a distance of 2ρ from $\partial\mathcal{R}_1$.) It suffices to show that

(5.27) $$x = X(\sigma, \tau) \in \mathcal{R}_2$$

if $(\sigma, \tau) \in \mathcal{S}^*_{\sigma_3}$. For if (5.27) holds, then by Theorem 5.1, $(\sigma, \tau) \in \mathcal{S}_2$; and we may conclude that $\mathcal{S}^*_{\sigma_3} \subseteq \mathcal{S}_2$. If $(\sigma, \tau) \in \mathcal{S}^*_{\sigma_3}$, then by (5.5) $|x - \hat{x}| = |X(\sigma, \tau) - F(\sigma, \tau)| = |R(\sigma, \tau)| \leq \rho$. ($\hat{x} = F(\sigma, \tau)$.) Since $\hat{x}$ lies in the region bounded by the surface

$$\{x | x = F(\sigma_3, \tau) ; \tau \in D''\} ,$$

and the surface

$$\{x | x = F(\sigma, \tau) , (\sigma, \tau) \in [\sigma_3, \infty) \times \partial D''\}$$

it follows that $x \in \mathcal{R}_2$, by the definition of $\mathcal{R}_2$; see Fig. 13. Thus we have proved

<u>Corollary</u> 5.2. <u>Under the hypotheses of Theorem</u> 5.1, $(\sigma, \tau) \to X(\sigma, \tau)$

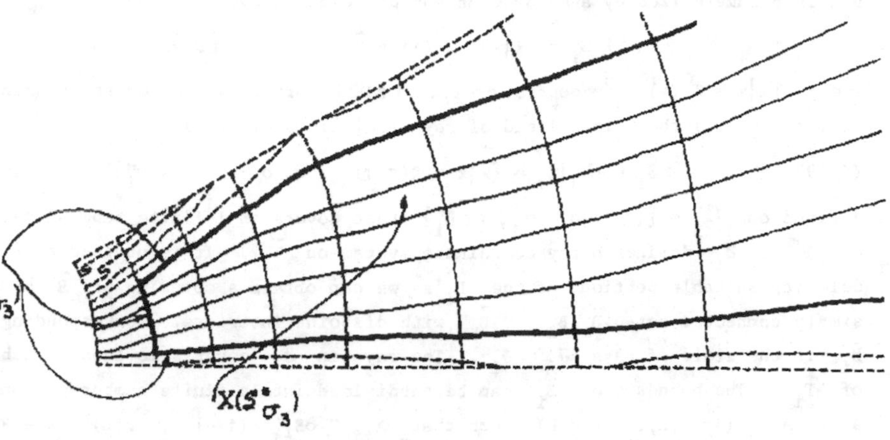

Figure 13

is a one to one mapping from $S^*_{\sigma_3}$ onto $\overline{T} = X(S^*_{\sigma_3})$. The tubular region T is bounded by the surface

$$\partial T - S = \{x | x = X(\sigma,\tau), (\sigma,\tau) \in [\sigma_3,\infty) \times \partial D''\} ,$$

which is a 2 parameter family of rays, and the wave front

$$S = \{x | x = X(\sigma_3,\tau), \tau \in D''\} .$$

The family of rays

$$\mathfrak{F} = [C_\tau | C_\tau = \{x | x = X(\sigma,\tau), \sigma \geq \sigma_3\} , \tau \in D'']$$

forms a field on $\overline{T}$, and defines a ray coordinate system on $\mathcal{R}_3 = \overline{T}$. Therefore T is an infinitely long ray cylinder. Every wave front

$$\Sigma_\sigma = \{x | x = X(\sigma,\tau) ; \sigma \geq \sigma_3 , \tau \in \partial D''\}$$

is locally convex relative to $n^{\frac{1}{2}}$.

For the definition of locally convex relative to $n^{\frac{1}{2}}$ see Definition 2.2 in Section 2. We remark that if ρ is chosen to be suitably small by choosing either σ_1 or h_0 sufficiently large, the regions $\mathcal{R}_i$ $(i = 1,2,3)$ will be well-defined with open interiors as illustrated in Figs. 10 - 13.

We conclude this Section by proving that $n^{\frac{1}{2}}$ gives rise to a field of rays $\tilde{\mathfrak{F}}$ on a full neighborhood $\mathfrak{N}$ of infinity in $\mathbb{R}^3$ $(\mathbb{R}^2)$; any orthogonal surface of this ray field is by Definition 2.3, convex relative to $n^{\frac{1}{2}}$.

Again let

$$\tilde{B} = \{x | x = \tilde{X}^0(\tau') , \tau' \in [\tau_1,\tau_2)\} \tag{5.28}$$

be a smooth closed convex surface. By virtue of Corollary 5.2 there is a finite covering of $\tilde{B}$ by closed, simply connected sets $\hat{B}_1, \hat{B}_2, \ldots, \hat{B}_k$ such that (i) each

$\hat{B}_i$ is parameterized by arcs of constant principle curvature; i.e.,

$$\hat{B}_i = \{x \mid x = X^0(\tau) = \tilde{X}^0(\tau'(\tau))\ ;\ \tau \in D_i''\}$$

where $\{x \mid x = X^0(\tau)\ ,\ \tau^p = \text{const.}, p = 1,2\ ,\ \tau \in D''\}$ are curves of constant principle curvature, (ii) there is a field of rays (solutions of 2.18)

$$(5.29) \qquad \mathfrak{F}_i = [C_\tau \mid C_\tau = \{x \mid x = X(\sigma,\tau) \quad \text{for} \quad \sigma \geq \sigma_3\ ,\ \tau \in D_i''\}]$$

defined on $\mathcal{S}^{*i}_{\sigma_3} = \{(\sigma,\tau) \mid \sigma \geq \sigma_3\ ,\ \tau \in D_i''\}$ that covers an infinite tubular region $\overline{T}_i$ of $\mathbb{R}^m$. $\mathfrak{F}_i$ defines a ray coordinate system on $\overline{T}_i$. It is obvious that by deleting suitable portions of the $\hat{B}_i$'s we can obtain a covering of $\tilde{B}$ by closed simply connected sets $\tilde{B}_1, \tilde{B}_2, \ldots, \tilde{B}_K$ with disjoint interiors. Corresponding to each $\tilde{B}_i$ is the field of rays $\tilde{\mathfrak{F}}_i \subseteq \mathfrak{F}_i$. The rays in $\tilde{\mathfrak{F}}_i$ cover some tubular subset $\overline{\tilde{T}}_i$ of T_i . The boundary of $\tilde{B}_i$ can be subdivided into a finite number of connected arcs A_{ij} $(j = 1,2,\ldots,L < K)$ such that $A_{ij} \subseteq \partial\tilde{B}_{i_j}$ $(j = 1,2,\ldots,L)$; see Fig. 14.

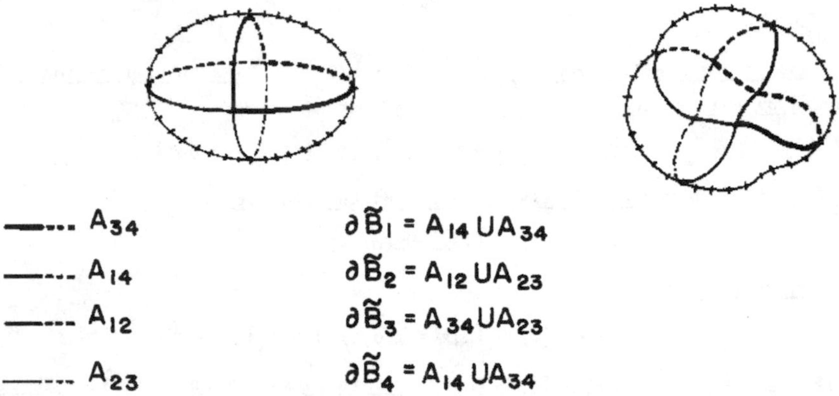

Figure 14

Let $\tilde{\mathfrak{F}}_{ii_j}$ be the subfamily of rays in $\tilde{\mathfrak{F}}_i$ that correspond to points in A_{ij} . Let $\tilde{\mathfrak{F}}_{i_j i}$ be the subfamily of rays in $\tilde{\mathfrak{F}}_{i_j}$ that correspond to points on A_{ij} .

Clearly $\tilde{\mathfrak{F}}_{ii_j} = \tilde{\mathfrak{F}}_{i_j i}$ since to each point $x = X^0(\tau)$ in A_{ij} there corresponds a unique solution $x = X(\sigma,\tau)$ of (2.18) for all $\sigma \geq \sigma_3$. We conclude that

<u>Corollary</u> 5.3. <u>Under the hypotheses of Theorem</u> 5.1, <u>if</u>

$$\tilde{B} = \{x \mid x = \tilde{X}^0(\tau')\ ,\ \tau' \in [\tau_1,\tau_2)\}$$

<u>is a closed convex surface</u>, <u>then</u>

$$\tilde{\mathfrak{F}} = \bigcup_{i=1}^{K} \tilde{\mathfrak{F}}_i$$

<u>is a field on the neighborhood of infinity</u>

$$\eta = \overline{\bigcup_{i=1}^{K} \tilde{T}_i}\ .$$

Furthermore,

$$\tilde{\mathfrak{F}} = [C_{\tau'} | C_{\tau'} = \{x = \tilde{X}(\sigma,\tau') \,,\, \sigma \geq \sigma_3\} \,,\, \tau' \in [\tau_1,\tau_2)] \tag{5.30}$$

where

$$\tilde{X}(\sigma,\tau') = \tilde{F}(\sigma,\tau') + \tilde{\mathfrak{F}}(\tilde{X}(\cdot,\tau') \,;\, \sigma,\tau') \,.$$

If $\tilde{X}^0(\tau') \in \tilde{B}_i$, *then* $\tilde{X}^0(\tau'(\tau)) = X^0(\tau)$ *and* $\tilde{X}(\sigma,\tau'(\tau)) = X(\sigma,\tau)$, *for all* τ *such that* $X^0(\tau) \in \tilde{B}_i$. *Since solutions of* (2.18) *satisfy* (2.8),

$$n(\tilde{X}(\sigma,\tau')) = \frac{1}{|\tilde{X}_\sigma(\sigma,\tau')|^2}$$

for all $(\sigma,\tau') \in [\sigma_3,\infty) \times [\tau_1,\tau_2)$; *and every closed orthogonal surface*

$$\Sigma_\sigma = \{x | x = \tilde{X}(\sigma,\tau') \,;\, \sigma = \text{fixed} \,,\, \sigma \geq \sigma_3 \,,\, \tau' \in [\tau_1,\tau_2)\}$$

is convex relative to $n^{\frac{1}{2}}$. *In particular* $\Sigma_0 = \partial\mathfrak{h}$ *is convex relative to* $n^{\frac{1}{2}}$.

The essence of Corollary 5.3 is that the ray tubes $\bar{\bar{T}}_i$ corresponding to the covering $\tilde{B}_1,\tilde{B}_2,\ldots,\tilde{B}_K$ of the closed convex surface $\tilde{B}$ fit together to form a ray field on a neighborhood $\mathfrak{h}$ of infinity in R^3 .

6. First derivatives of X and the Jacobian

In this Section we outline the proof of

Theorem 6.1. Under the hypotheses of Theorem 5.1, the solution X *of the ray equations* (2.18), *whose existence is guaranteed by Proposition* 4.1, *has continuous first derivatives* X_σ , X_{τ^i} $(i = 1, m-1)$, *and uniformly on* $S^*_{\sigma_3}$

$$X_\sigma = \nu(\tau) + \mathcal{O}(\sigma^{-2}) \,, \tag{6.1a}$$

$$X_{\tau^i} = \sigma\nu_{\tau^i} + X^0_{\tau^i} + \mathcal{O}(\sigma^{-1}) \,. \tag{6.1b}$$

Moreover, the Jacobian $\partial X/\partial(\sigma,\tau)$ *has the form*

$$J(\sigma,\tau) = J_f(\sigma,\tau) + \mathcal{O}(\sigma) \tag{6.2}$$

uniformly on $S^*_{\sigma_3}$, *where* J_f *is the Jacobian of the map* $x = \nu(\tau) + X^0(\tau)$. *Here* J_f *is given by formula* (6.2) *in* [1], *and*

$$c_1\sigma^2 \leq J_f(\sigma,\tau) \leq c_2\sigma^2 \qquad (\sigma,\tau) \in S^*_{\sigma_3} \tag{6.3}$$

for some positive constants c_1 *and* c_2 .

The proof presented below is an outline of the main steps and conclusions only.

Proof. That X_σ exists, is continuous and satisfies the condition (6.1a) is an easy consequence of (2.18); one simply differentiates the right-hand side of (2.18) with respect to σ and uses the hypotheses (2.20)-(2.23) for $|p| \leq 2$ and the estimate (5.20). To obtain the existence and asymptotic behavior of the derivatives X_{τ^i} is a more delicate matter since the integrands in (2.18) depend upon τ and since the range of integration is infinite.

To show existence and continuity of the X_{τ^i} one argues analogously to [2, Chapter I] by studying the difference

$$\hat{x}_i - \frac{\Delta x}{\Delta \tau^i}$$

where $\hat{x}_i$ is the solution of the linear integral equation obtained by formally differentiating (2.18) with respect to τ^i and replacing X_{τ^i} by $\hat{x}_i$. A slight difficulty is caused by the integration in (2.18) being taken over an unbounded range, but the kernels in (2.18) decay fast enough that the argument is successful. The conclusion is that $X_{\tau^i} = \hat{x}_i$.

Once existence and continuity of the X_{τ^i} are established, it is a straightforward matter to estimate them using (5.20) and the hypotheses (2.20)-(2.23) for $|p| \leq 2$, since the integral equations for the X_{τ^i} are linear. Let

$$y_{\tau^i} = X_{\tau^i} - F_{\tau^i}(\sigma,\tau) .$$

Then one can show that

$$|y_{\tau^i}| \leq \text{Const.} \int_\sigma^\infty t^{-2} |y_{\tau^i}(t,\tau)| dt + \gamma(\sigma) , \tag{6.4}$$

where γ is a smooth function and $\gamma(\sigma) = \mathcal{O}(\sigma^{-1})$ as $\sigma \to \infty$. Let η denote the integrand in (6.4). Then (6.4) is a differential inequality for η, from which it follows by integration that

$$\begin{aligned} |y_{\tau^i}| &\leq \text{Const.}(\sigma^{-1} + 1)\gamma(\sigma) , \\ &= \mathcal{O}(\sigma^{-1}) . \end{aligned}$$

This proves (6.1b).

The results (6.2) and (6.3) can be obtained by straightforward computation and estimation using the previous results (6.1a) and (6.1b) in (2.11).

This completes our discussion of the proof of Theorem 6.1.

7. Higher Derivatives of X

In order to establish a rigorous approximate solution of the scattering problem (U) of the Introduction, it is necessary to establish the smoothness and asymptotic behavior of the higher order derivatives of X, the solution of (2.18) that we have been studying. The order of the highest derivative needed depends upon the degree of accuracy to which one wishes to satisfy the problem (U). In particular, if the approximation to the solution is to be accurate to within

$$\mathcal{O}\left(\lambda^{-N+\frac{1}{2}(1+m)} |x|^{\frac{1}{2}(1-m)}\right) \qquad (x \in \overline{V}) ,$$

then continuity and the asymptotic behavior for large σ of the first $2N+5$ derivatives of X are sufficient.

The analysis of these higher derivatives is primarily an accounting problem. Hence, we only describe the analysis that leads to the results we have obtained; and we omit all the computational details, although we have executed them ourselves.

To describe this analysis we divide the derivatives into two classes: pure τ derivatives and all others. We begin with the scalar equations for the components of X, which are obvious from (2.18) and which we rewrite in abbreviated notation as

$$X^i = v^i + \nu^i \int_\sigma^\infty F^1(X)ds + \int_\sigma^\infty F^2(X)\int_\sigma^\infty G^i(X)dt\,ds \quad . \tag{7.1}$$

We proceed by induction. We have completed the first induction step in proving Theorem 6.1.

To derive the integral equations for

$$D^{(p,q)}X^i \qquad \left(D^{(p,q)} \triangleq \partial^p_{\tau_1}\partial^q_{\tau_2}\right)$$

we use Leibnitz' rule and the chain rule. Eventually, in the case $m=3$ we obtain a system of three linear Volterra equations. The kernel of this system is a function of the X^i only. It is the same kernel that occurs in the study of X_{τ^i}. The inhomogeneous term involves sums of products of lower order derivatives $D^{(p',q')}X^i$ $(0 \le p' \le p, 0 \le q' \le q, p'+q' < p+q)$ with functions of X. Each summand is estimated using the induction hypothesis and the assumed behavior (2.20)-(2.23) of n and its derivatives. We thus obtain a system of equations of the form

$$\begin{aligned} D^{(p,q)}X^i &= D^{(p,q)}v^i + \nu^i \int_\sigma^\infty \nabla F^1 \cdot D^{(p,q)}X\,ds \\ &+ \int_\sigma^\infty\int_s^\infty [F^2\nabla G^i + G^i\nabla F^2]\cdot D^{(p,q)}X\,dt\,ds + \mathcal{O}(\sigma^{-1}) \quad . \end{aligned} \tag{7.2}$$

As in the case of the $X^i_{\tau^j}$, we need only consider (7.2) rewritten for the derivatives of the components of $y = X - \nu\sigma - X^0$. The kernels in these new equations have known orders of magnitude in σ. A standard differential inequality argument yields a final estimate for $D^{(p,q)}y^i$, just as in the case of the first derivatives.

Existence of the $D^{(p,q)}X$ is obtained by induction through the argument used in the case $p+q = 1$.

It remains to consider

$$u_m = \partial^m_\sigma(D^{(p,q)}X) \quad .$$

The cases $m=1$ and $m>1$ are slightly different (here m is <u>not</u> the dimension of R^m):

$$u^i_1 = D^{(p,q)}[\nu^i F^1(X) + F^2(X)\int_\sigma^\infty G^i(X)ds + \nu^i] \quad , \tag{7.3}$$

and for $m > 1$,

$$u^i_m = \partial^{m-1}_\sigma D^{(p,q)}[\nu^i F^1(X) + F^2(X)\int_\sigma^\infty G^i(X)ds] \quad . \tag{7.4}$$

The behavior for $\sigma \to \infty$ of the terms on the right-hand side of (7.3) has already been determined. Thus we conclude that

(7.5) $$u_1^i = D^{(p,q)}\nu^i + \mathcal{O}(\sigma^{-2}) \quad .$$

It is clear from (7.4) that the behavior of u_m^i depends on that of $u_{m'-1}^i$ $(1 \leq m' < m)$ and the derivatives of F^1, F^2 and G with respect to X^i. An induction argument, use of Leibnitz' rule and the chain rule lead to the result that for $m > 1$

(7.6) $$u_m^i = \mathcal{O}(\sigma^{-1-m}) \quad .$$

The final conclusions resulting from the previous argument are stated as part of the main theorem in the following Section.

We also want to show that if a ray system is given with the asymptotic behavior we have established in Section 6 and thus far in this Section (this is summarized in (8.1) below), then that ray system defines an index of refraction $n^{\frac{1}{2}}(x)$ that satisfies the hypotheses (2.20)-(2.23). We shall not do this in detail for the calculuations are tedious. We merely indicate how to carry out the analysis, which is straightforward.

By the assumption that a ray system is defined on a region $\overline{T}$ it follows that $X(\sigma(x),\tau(x))$ is given on $\overline{T}$ and that X has the smoothness and asymptotic properties (8.1) in σ and τ. Given X we define $n(x)$ by (2.8). For convenience we write

(7.7) $$n(x) \equiv g(\sigma(x),\tau(x)) \equiv |X_\sigma(\sigma(x),\tau(x))|^{-2} \quad .$$

Note that

$$x \equiv X(\sigma(x),\tau(x)) \qquad (x \in \overline{T}) \quad .$$

It immediately follows from (7.7) and (8.1) that n is continuous in x if σ and τ are continuous functions of x. The continuity of σ and τ is a consequence of the smoothness of X with respect to σ and τ, the positivity of the Jacobian of X, and the Implicit Function Theorem.

In further discussing the smoothness and asymptotic behavior of n and its derivatives we consider the case $x \in \mathbf{R}^3$ only. First we compute the asymptotic behavior of σ with respect to x. Since, by hypothesis,

$$x = X(\sigma,\tau) = \nu(\tau)\sigma + X^0(\tau) + \mathcal{O}(\sigma^{-1})$$

uniformly on $X^{-1}(\overline{T})$, it follows directly that

(7.8) $$|x|^2 = \sigma^2 + \mathcal{O}(1)\sigma \quad .$$

Hence, if $|x| \to \infty$, then $\sigma \to \infty$. Treating (7.8) as a quadratic equation for σ, we find that

(7.9) $$\sigma = |x| + \mathcal{O}(1) \qquad \text{on } \overline{T} \, .$$

Thus the $\mathcal{O}(\sigma^{-k})$-terms in (8.1) are $\mathcal{O}(|x|^{-k})$. It now follows from (7.7) and (8.1) that

(7.10) $$n(x) = 1 + \mathcal{O}(|x|^{-2}) \qquad \text{on } \overline{T} \quad .$$

We next consider the first derivatives of $n(x)$. By the chain rule,

$$\partial_\sigma g = \nabla n \cdot X_\sigma ,$$
$$\partial_{\tau^i} g = \nabla n \cdot X_{\tau^i} \qquad (i=1,2) .$$

This is a linear system for the first (Cartesian) derivatives of n, which we solve by Cramer's rule. For example,

$$(7.11)\qquad n_{x^1} = \frac{\begin{vmatrix} \partial_\sigma g & X^2_\sigma & X^3_\sigma \\ \partial_{\tau^1} g & X^2_{\tau^1} & X^3_{\tau^1} \\ \partial_{\tau^2} g & X^2_{\tau^2} & X^3_{\tau^2} \end{vmatrix}}{J} .$$

The continuity of n_{x^1} in x is a consequence of this formula, (8.1), and the continuity of σ and τ as functions of x. Further, the asymptotic behavior of the right-hand side of (7.11) with respect to σ is given by (8.1). Thus using (8.1), (7.9), and the identities

$$\nu \cdot \nu_{\tau^i} \equiv 0 \qquad (i=1,2) ,$$

we conclude that

$$n_{x^1} = \mathcal{O}(|x|^{-3}) \quad \text{on } \overline{T} .$$

The smoothness and asymptotic behavior of the remaining first derivatives of n is obtained similarly.

The above argument can be extended by induction to obtain the continuity and asymptotic behavior of the higher derivatives of n. We do not explicitly carry out the induction. Rather we illustrate the argument for the second derivatives $n_{x^1x^j}$ $(j=1,2,3)$.

Differentiating both sides of (7.11) successively with respect to σ, τ^1, and τ^2, we obtain the identities

$$(7.12)\qquad \begin{aligned} \nabla n_{x^1} \cdot X_\sigma &= -\frac{J_\sigma}{J} n_{x^1} + J^{-1}\Gamma_\sigma \equiv \psi^1 \\ \nabla n_{x^1} \cdot X_{\tau^i} &= -\frac{J_{\tau^i}}{J} n_{x^1} + J^{-1}\Gamma_{\tau^i} \equiv \psi^{i+1} \qquad (i=1,2) , \end{aligned}$$

where Γ is the determinant exhibited in (7.11). The determinants Γ_σ and Γ_{τ^i} are functions of the second derivatives of g and the first and second derivatives of X^i $(i=1,2,3)$. The second derivatives of g are functions of the first, second and third derivatives of X^i $(i=1,2,3)$. The system (7.12) is a linear system for the $n_{x^1x^j}$ $(j=1,2,3)$. Solving it, we find that

$$(7.13)\qquad n_{x^1x^j} = \frac{\Delta^j}{J} \qquad (j=1,2,3),$$

where Δ^j is the determinant obtained by replacing the j-th column of the determinant J by $\mathrm{col}(\psi^1,\psi^2,\psi^3)$.

As was the case before with respect to (7.11) the behavior of the right-hand side of (7.13) can be found from (8.1), (7.9), the identities $\nu \cdot \nu_{\tau^i} \equiv 0$ $(i=1,2)$, and the continuity of σ and τ with respect to x. We conclude that the $n_{x^1x^j}$ $(j=1,2,3)$ are continuous on $\overline{T}$ and that

$$n_{x^1x^j} = \mathcal{O}(|x|^{-4}) \qquad (x \in \overline{T}) .$$

8. The Main Theorem

The hypotheses of Theorem 2.7 are implied by Corollary 5.2, Theorem 6.1, and the C^{2N+5}-smoothness of X (together with the asymptotic behavior of X and its derivatives) established by the arguments in Section 7. Thus the results we have obtained in Sections 2 - 7 imply that X determines a ray system. We summarize this and the results of the previous Sections of this Chapter in one main theorem:

Theorem 8.1. Let $n^{\frac{1}{2}}$ be an index of refraction defined on $\mathbb{R}^m$ $(m=2$ or $3)$ and having the smoothness and asymptotic properties (2.20)-(2.23), N being a positive integer. Further,

A) Suppose a field $\hat{\mathfrak{F}}$ of straight-line rays is chosen with the rays in $\hat{\mathfrak{F}}$ emanating normally from a locally convex patch $\hat{B}$ on a surface (arc) in $\mathbb{R}^m$. Then, corresponding to $n^{\frac{1}{2}}$, N, and $\hat{B}$, there exists a tube $\overline{T}$, extending to ∞, on which a field $\mathfrak{F}$ of rays is defined, with the rays in $\mathfrak{F}$ emanating normally from a patch that is locally convex relative to $n^{\frac{1}{2}}$. The rays in $\mathfrak{F}$ are asymptotic to those in $\hat{\mathfrak{F}}$ at ∞. The size of T depends on n, and the rate at which $n \to 1$ as $|x| \to \infty$. Further, on $\overline{T}$ the solution X of (2.18) defines an orthogonal ray coordinate system, and X has the following smoothness and asymptotic properties on $\overline{T}$:

$$(8.1)\qquad \begin{aligned} &X \in C^{2N+5}, \quad X(\sigma,\tau) = \nu(\tau)\sigma + X^0(\tau) + \mathcal{O}(\sigma^{-1}), \\ &X_\sigma(\sigma,\tau) = \nu(\tau) + \mathcal{O}(\sigma^{-2}), \ X_{\tau^i}(\sigma,\tau) = \sigma\nu_{\tau^i}(\tau) + X^0_{\tau^i}(\tau) + \mathcal{O}(\sigma^{-1}) \quad (i=1,m-1), \\ &D^{(p,q,r)}X = \mathcal{O}(\sigma^{1-p}) \quad (0 \le p+q+r \le 2N+4), \\ &D^{(p,q,r)}X = \mathcal{O}(\sigma^{-1-p}) \quad (0 \le p+q+r \le 2N+5,\ p \ge 2), \end{aligned}$$

where $x = X^0(\tau)$ on $\hat{B}$ and $D^{(p,q,r)} = \partial^p_\sigma \partial^q_{\tau^1} \partial^r_{\tau^{m-1}}$. The estimates in (8.1) are each uniform on $\overline{T}$;

and

B) Suppose a field $\hat{\mathfrak{F}}$ of straight-line rays is chosen with the rays in $\hat{\mathfrak{F}}$ emanating normally from a convex body $\tilde{B}$ in $\mathbb{R}^m$. Then, corresponding to $n^{\frac{1}{2}}$, N, and $\tilde{B}$, there exists a neighborhood $\mathfrak{n}$ of ∞ in $\mathbb{R}^m$ on which a field $\tilde{\mathfrak{F}}$ of rays is defined, with the rays in $\tilde{\mathfrak{F}}$ emanating normally from a smoothly embedded $(m-1)$-sphere that is convex relative to $n^{\frac{1}{2}}$. The rays in $\tilde{\mathfrak{F}}$ are asymptotic to those in $\hat{\mathfrak{F}}$ at ∞. The size of $\mathfrak{n}$ depends on n and the rate at which $n \to 1$ as $|x| \to \infty$. Moreover, $\mathfrak{n}$ is covered by a finite number of tubes $\overline{T}_i$, extending to ∞, such that the solution X^i of (2.18) on $\overline{T}_i$ defines an orthogonal ray coordinate system on $\overline{T}_i$ and X_i has the smoothness and asymptotic properties (8.1) on $\overline{T}_i$ with X re-

<u>placed by</u> X^i .

C) <u>Conversely, if a function</u> X^i , <u>satisfying the conditions</u> (8.1), <u>that defines an orthogonal ray coordinate system on a tube</u> $\overline{T}_i$, <u>extending to</u> ∞ , <u>is given, then that ray system gives rise to an index of refraction</u> $[n'(x)]^{\frac{1}{2}}$ <u>on</u> $\overline{T}_i$ <u>having the properties</u> (2.20) - (2.23), <u>and</u>

$$[n'(X^i)]^{\frac{1}{2}} = [X^i_\sigma \cdot X^i_\sigma]^{\frac{1}{2}} \quad .$$

<u>Moreover, if that ray system came from a given index of refraction</u> $n^{\frac{1}{2}}$ <u>with the properties</u> (2.20) - (2.23), <u>then</u> $n' = n$ <u>on</u> $\overline{T}_i$. <u>Thus the index of refraction on on</u> $\overline{T}_i$ <u>is uniquely determined by</u> X^i .

References for Chapter 2

1. C. O. Bloom and N. D. Kazarinoff, Local energy decay for a class of nonstar-shaped bodies, Archive for Rat. Mech. and Anal., 55 (1974), 73-85.
2. E. Coddington and N. Levinson, <u>Theory of Ordinary Differential Equations</u>, McGraw-Hill, N. Y., 1955.
3. D. S. Jones, High-frequency refraction and diffraction in general media, Philos. Trans. Roy. Soc. London Ser. A 255 (1962/63), 341-387.
4. M. Klein and I. W. Kay, <u>Electromagnetic Theory and Geometrical Optics</u>, Wiley (Interscience), N.Y., 1965.
5. R. K. Luneberg, <u>Mathematical Theory of Optics</u>, U. of Calif. Press, Berkeley, 1964.

CHAPTER 3

A UNIFORM APPROXIMATION TO THE SOLUTION OF URSELL'S RADIATING BODY PROBLEM

1. Introduction

In this Chapter we construct a rigorous N^{th}-order approximate solution $u_N(x,\lambda)$ to the exterior radiating body problem (Problem (P)):

$$Lu \overset{d}{=} \Delta u + \lambda^2 n(x)u = f(x) \qquad (x \in V) , \tag{1.1}$$

$$u(x) = u_0(x) \qquad (x \in \partial V) , \tag{1.2}$$

$$\lim_{R\to\infty} \int_{|x|=R} r|u_r - i\lambda u + \frac{m-1}{2r} u|^2 = 0 , \tag{1.3}$$

where $x \in R^m$ $(m = 2 \text{ or } 3)$ and $r = |x|$. We consider scattering obstacles of finite cross-section with boundaries ∂V consisting of a finite number of smooth patches S_i joined together so that

$$\partial V = Q \cup (\cup S_i)$$

is a smooth embedding of an $(m-1)$-sphere in R^m. Each patch S_i is required to be locally convex relative to the given index of refraction $n^{\frac{1}{2}}$; see Definition 2.2 of Chapter 2. In particular this includes the case where ∂V is convex relative to $n^{\frac{1}{2}}$; take $Q = \emptyset$ and $S_1 = \partial V$; see Definition 2.3 of Chapter 2.

The region V in which the solution of (1.1) - (1.3) is sought is the <u>exterior</u> of ∂V. We further assume that no optical path emanating normally from a patch S_i intersects $\partial V - S_i$ (no reflections). If ∂V can be illuminated from the exterior (see Definition 2.1 of Section 2, Chapter 1), we can apply the a priori estimates derived in Chapter 1 to $|u - u_N|$ and conclude that u_N is an asymptotic expansion of the exact solution u.

Although we impose the restriction that rays emanating normally from each patch S_i on ∂V not intersect $\partial V - S_i$, <u>rays emanating from different patches may intersect in</u> V.

The support of the source function f in (1.1) is required to lie in $V \cup (\cup S_i)$. Furthermore, the support of the boundary data u_0 is assumed to lie in $\cup S_i$, and f and u_0 should be sufficiently smooth. The precise statement of our assumptions and results is:

<u>Theorem</u> 1.1. <u>Suppose that a positive integer</u> N <u>is given, that</u> $n^{\frac{1}{2}}$ <u>is a given index of refraction on</u> R^m $(m = 2$ <u>or</u> $3)$, <u>and that</u> ∂V <u>is a smooth</u> C^{2N+5} <u>scattering obstacle that can be illuminated from the exterior with</u>

$$\partial V = (Q \cup (\cup S_i)) \subset R^m \qquad (m = 2 \text{ or } 3) , \tag{1.4}$$

where the S_i *are smooth patches each of which is locally convex relative to* $n^{\frac{1}{2}}$. *Suppose further that the following hypotheses are satisfied:* (1) *no ray emanating normally from any patch* S_i *intersects* $\partial V - S_i$, (2) *the index of refraction* $n^{\frac{1}{2}}$ *satisfies the hypotheses made in Chapters* 1 *and* 2, (3) $f \in C^{2N}(V \cup (\cup S_i))$, $D^p f = \mathcal{O}(r^{-|p|})$ *for* $|p| \le N$ *on* $\bar{V}$ *if* N *is even and for* $|p| \le N+1$ *if* N *is odd,* (4) $\int_V r^2 |f|^2 < \infty$, *and* (5) $u_0 \in C_0^{2N+4}(\cup S_i)$. *Then if* u *is the solution of* (P),

$$(1.5) \qquad |u - u_N| = \mathcal{O}(\lambda^{-N+\frac{1}{2}(1+m)} r^{\frac{1}{2}(1-m)}) \qquad (m = 2 \text{ or } 3,\ \lambda \to \infty)$$

uniformly in x *on* V, *where* u_N *is the approximate solution whose terms are defined by* (0.40) - (0.50) *and* (0.54) *of Chapter* 0.

In view of the a priori estimate (1.8.9) established in Chapter 1, to prove Theorem 1.1 it suffices to construct a function $u_N(x,\lambda)$ such that

$$(1.6) \qquad Lu_N = f(x) + f_1(x,\lambda) \qquad (x \in V),$$

where

$$(1.7) \qquad f_1(x,\lambda) = \mathcal{O}(\lambda^{-N} r^{-(m+3)/2}) \qquad (\lambda \to \infty)$$

uniformly in x on $\bar{V}$, and where u_N satisfies the boundary condition (1.2) and the radiation condition (1.3). (Note that if each of two functions satisfies the radiation condition, then their sum does also.) Application of (1.8.9) to $u - u_N$ then yields the desired result.

Before outlining the organization of the proof of Theorem 1.1, we emphasize that this theorem applies to situations where there is weak focussing of rays. For example, suppose that S_1 and S_2 are two patches on ∂V and $u_0 = u_{01} + u_{02}$, where u_{0i} $(i = 1,2)$ is smooth and is supported on S_i. For simplicity let $f = 0$. Finally, assume that the fields of rays emanating normally from S_1 and S_2 intersect in V. Then the approximate solution u_N can be constructed as follows. Let u_{Ni} $(i = 1,2)$ be the approximate asymptotic solutions corresponding to the data u_{0i}, the u_{Ni} being constructed according to the prescription given in Chapter 0 (see (0.40)-(0.81)). Then Theorem 1.1 applies to

$$u_N = u_{N1} + u_{N2};$$

and u_N is a rigorous asymptotic approximation to the solution u of the radiation-scattering Problem P.

Note that if $u_0 = 0$ and f has compact support with

$$(\operatorname{supp} f) \cap \partial V = \emptyset,$$

then only smoothness of f and n are required to establish the conclusions of Theorem 1.1 in addition to the hypotheses required for the a priori estimates of Chapter 1 to hold. In this case the approximate solution u_N is zero outside $\operatorname{supp} f$. Moreover, if f and n are C^∞ on $\operatorname{supp} f$, this implies that the exact solution u

is smaller than any inverse power of λ $(\lambda \to \infty)$ off the support of f (provided ∂V can be illuminated from the exterior).

There is a less obvious case for which the a priori estimates of Chapter 1 and an Ansatz for an approximate solution would yield a rigorous approximation to u. This is the case where f is a point-source, represented by a δ-function, $u_0 \equiv 0$, and u is the Green's function $G(x,x')$. If an Ansatz could be found in this case, one could apply the a priori estimates of Chapter 1 to $G - u_N$. It should be possible to construct an approximate solution u_N in this case by adapting the construction used by D. Ludwig and C. Morawetz in [1] to the case of inhomogeneous media. In any case where an approximate solution u_N can be constructed in the exterior of an obstacle that can be illuminated from the exterior our a priori estimate (1.8.9) can be applied to prove the u_N is a rigorous asymptotic approximation to u. We thank Professor J. B. Keller for these observations.

We also believe that the analysis presented in these Lecture Notes can be extended to cases where rays emanating from a locally convex patch S_i are reflected a finite number of times from other patches, including possibly S_i, before reaching infinity.

The remainder of this Chapter is organized as follows. We discuss the formal Ansatz in Section 2. We establish the smoothness and asymptotic behavior of the terms in this Ansatz in Section 3, using the results from Chapter 2. In Section 4 we show that the approximate solution u_N satisfies the radiation condition (1.3). Sections 2 - 4 deal with obstacles that are convex relative to $n^{\frac{1}{2}}$. We complete the proof of Theorem 1.1 for more general obstacles in Section 5.

In Section 6 we consider more general radiation problems where u satisfies a boundary condition of the form

$$\alpha(x,\lambda)u + \beta(x,\lambda)\nu^*(x) \cdot \nabla u = g(x) \qquad (x \in \partial V) \tag{1.8}$$

in addition to equation (1.1) and the radiation condition (1.3); see the beginning of Chapter 0 for notation. We present an algorithm for constructing an approximate solution u_N^* of (1.1), (1.8), and (1.3) under the hypotheses that $\alpha(x,\lambda)$ and $\beta(x,\lambda)$ have asymptotic expansions in integral powers of λ^{-1}. The case where

$$\alpha(x,\lambda) = \alpha_1(x) - i\lambda c\alpha_2(x)$$

and $\beta(x,\lambda)$ is independent of λ is of special interest since the solution u is in this case the amplitude of the steady state solution of (0.7) - (0.9) of Chapter 0.

To show that this u_N^* is an asymptotic expansion of u it is necessary to generalize the a priori estimates of Chapter 1 to the more general boundary condition (1.8). There are formidable obstacles to achieving this in the case $\beta(x,\lambda) \not\equiv 0$, although progress has been made in this direction for the case $n \equiv 1$ and ∂V convex; see [2].

On the other hand if $\beta \equiv 0$, it should be possible to generalize the a priori estimates of Chapter 1 to boundaries that do not trap rays. This problem has been treated for the case $n \equiv 1$ in [2].

2. The Ansatz

In this and in the following sections, except Sections 5 and 6, we assume that ∂V is convex relative to $n^{\frac{1}{2}}$. We now ask the reader to recall the formulae (0.40) - (0.46) of Chapter 0. It follows from Theorem 8.1 of Chapter 1 that the desired estimate 1.5 for $|u - u_N|$ will be established for V convex relative to $n^{\frac{1}{2}}$ with an f in (1.6) satisfying (1.7) provided: (1) we can show that there exist $C^{2(N-j)+2}(\overline{V})$-functions A^{j+1} $(j = 0,\ldots,N)$ and $C^{2(N-j)}(\overline{V})$-functions B^{2j} $(1 \leq 2j \leq N-2$ if N is even; $1 \leq 2j \leq N-1$ if N is odd) (and an appropriate χ) satisfying the conditions (0.42) - (0.45) of Chapter 0 on $\overline{V}$ and such that, as $r \to \infty$, ΔA^{N+1}, ΔB^k $(k = N-2, N-1)$ are each

$$\mathcal{O}(r^{-(m+3)/2}) ;$$

and (2) we can show that $u - u_N$ satisfies the radiation condition (1.3).

The radiation condition (1.3) will be satisfied by $u - u_N$ if both u and u_N satisfy it. It is satisfied by u, since, by hypothesis, u solves (1.1) - (1.3). It will be satisfied by u_N if

$$\lim_{R\to\infty} R\int_{|x|=R} |h|^2 = 0 , \tag{2.1}$$

where h is any one of the following functions

$$A_r + i\lambda(\chi_r - 1)A + \frac{m-1}{2r}A , \quad \text{where } A = \Sigma_1^{N+1}\lambda^{-n}A^n ,$$

$$B^0 , \quad B_r^{N'-2}\ \frac{m-1}{2r}B^{N'-2}, \qquad (N' = N+2,\ N \text{ odd}; \ N' = N,\ N \text{ even}) \tag{2.2}$$

and

$$B_r^{j-1} + \frac{m-1}{2r}B^{j-1} - iB^j \qquad (j = 1,\ldots,N'-2) .$$

To verify that the construction of u_N as prescribed in Chapter 0 can be carried out successfully and that it yields an approximation to the desired accuracy for u we next study the smoothness and far-field behavior of u_N and $L(u - u_N)$.

3. Analysis of the A^j and B^{j-2}

We assume now that $n^{\frac{1}{2}}$, the index of refraction, satisfies the hypotheses of Theorem 1.1, and that, by Theorem 8.1 of Chapter 2, it generates a ray system X on a neighborhood $\mathfrak{h}$ of infinity that contains $\overline{V}$ in its interior, where ∂V is convex relative to $n^{\frac{1}{2}}$, namely

$$\partial V = \{x \mid x = X(\sigma,\tau) ,\ \sigma = \sigma_3 > 0 ,\ \tau_1 \leq \tau < \tau_2\} . \tag{3.1}$$

It is here that the assumption $\partial V \in C^{2N+5}$ is made; in Theorem 8.1 of Chapter 2, in which the existence of a ray system corresponding to $n^{\frac{1}{2}}$ is asserted, the level surfaces σ = constant are furnished with C^{2N+5} smoothness. Note that with the notation adopted in (3.1) the function χ is described by

(3.2) $$\chi = \sigma(x) - \sigma_3 \quad \text{not} \quad \chi = \sigma(x) \ .$$

The analysis of the A^j and B^{j-2} requires smoothness and asymptotic properties of the ray system generated by the given $n^{\frac{1}{2}}$; and, ultimately then, all depends on the smoothness and asymptotic properties of n . Theorem 8.1 of Chapter 2 provides the desired connection between properties of n and of the ray field $\tilde{\mathfrak{F}}$ on $\overline{V}$ defined in (2.5.30).

The recursion formulas (0.45), (0.50) and (0.53) of Chapter 0 are the point of beginning of this analysis. Our general approach is as follows. We make an induction hypothesis on A^j and B^{j-2} for $j < N+1$, and we use it and the recursion formulas to establish asymptotic and smoothness properties of A^{N+1} and B^{N-1} . This involves considerable work.

The transport equations (0.43), (0.44) for the $A^j(x)$ hold on $\overline{V}$ as do the equations (0.45) for the $B^p(x)$. However, as noted in Chapter 0, in order to integrate the transport equations we introduce local ray coordinates $\sigma = \sigma(x)$, $\tau^i = \tau^i(x)$, where $\sigma(x)$, $\tau^i(x)$ are the unique functions satisfying

$$x = X_i(\sigma, \tau^i) \ .$$

The ray family $\mathfrak{F}_i$ (see 2.5.29) forms a field on an infinite tubular region T_i , and $\overline{\cup_i T_i} = \overline{V}$. Recall that the rays in the family $\mathfrak{F}_i$ correspond to the field of straight lines rays

$$\hat{\mathfrak{F}}_i = [\hat{C}_{\tau^i} | \hat{C}_{\tau^i} = \{x | x = \nu(\tau^i)\sigma + X^0(\tau^i) \ , \ \sigma \geq 0 \ , \ \tau^i \in D_i\}]$$

and that the rays in $\mathfrak{F}_i$ are asymptotic to the straight line rays in $\hat{\mathfrak{F}}_i$. Further, recall that the τ^i are chosen so that the curves on the reference patch $\{x | x = X^{0i}(\tau^i) \ , \ \tau^i \in D_i\}$, where one of the components of τ^i is constant, are curves of constant principal curvature. For further reference we define

$$\mathcal{S}_i = \{(\sigma, \tau) | \sigma \geq \sigma_3 \text{ and } \tau \in D_i\} \ .$$

We first compute the Laplace operator in local (σ, τ) coordinates described above and establish the asymptotic and smoothness properties of the coefficients and their derivatives in the Laplacian as functions of σ and τ . We shall outline the calculations for the case $m = 3$, which is the more difficult case.

In local (σ, τ)-coordinates corresponding to a typical ray tube $\overline{T}_i$ in $\overline{V}$,

(3.3) $$J\Delta u = (nJu_\sigma)_\sigma + \sum_{i=1}^{2} (H^{3-i,3-i} u_{\tau^i\tau^i} + E^{3-i} u_{\tau^i}) - 2H^{1,2} u_{\tau^1\tau^2} \ ,$$

where J is the Jacobian of the transformation defined by $X(\sigma, \tau)$,

(3.4) $$H^{i,j} = \frac{X_{\tau^i} \cdot X_{\tau^j}}{nJ} \quad (i, j = 1, 2) \ , \text{ and}$$

$$E^1 = H^{1,1}_{\tau^2} - H^{1,2}_{\tau^1} \ , \ E^2 = H^{2,2}_{\tau^1} - H^{1,2}_{\tau^2} \ .$$

We have chosen the local coordinates (σ,τ) so that the coefficients in (3.3) have no singularities and J is bounded away from zero. If we use (3.3) to compute ΔA^j and we integrate by parts to remove the σ derivatives from A^j under the integral sign in (0.54) of Chapter 0, we obtain the formula

$$a^{-1}A^{j+1} = \Gamma^j(\tau) + c(\sigma,\tau)A^j_\sigma + g(\sigma,\tau)A^j$$

(3.5)
$$+ \int_{\sigma_3}^{\sigma} [fA^j + e^\ell A^j_{\tau^\ell} + d^{\ell,p}A^j_{\tau^\ell\tau^p}]d\hat{\sigma}$$

$$(j = 0,\dots,N) ,$$

where repeated indices are to be summed from 1 to 2, and where

$$a = (nJ)^{-\frac{1}{2}} , \quad c = \frac{i}{2a} , \quad d = \frac{i}{2a}\partial_\sigma \ln(nJ) ,$$

(3.6)
$$f = c_{\sigma\sigma} - d_\sigma , \quad g = d - c_\sigma ,$$

$$d^{3-k,3-\ell} = \frac{ia}{2} H^{k,\ell}(-1)^{k+\ell} , \quad e^\ell = \frac{iaE^{3-\ell}}{2} \qquad (k,\ell = 1,2) ,$$

and

$$\Gamma^j = (1 - \delta_{j,0})B^{j-1}(X(\sigma_3,\tau)) - [cA^j_\sigma + gA^j]_{\sigma=\sigma_3} .$$

Note that

(3.7)
$$\Delta X = n\partial_\sigma \ln(nJ) .$$

Thus smoothness of n and J, which is established below, and the strict positive definiteness of n and J more than imply the necessary smoothness of χ (necessary for Lu_N to exist).

We shall use the notation

$$D^q = \partial_\sigma^{q_1}\partial_{\tau^1}^{q_2}\partial_{\tau^2}^{q_3} \quad \text{and} \quad |q| = \Sigma_1^3 q_i ,$$

where the q_i are nonnegative integers, throughout this Section. We wish to prove by induction on $|q|$ that for $m = 2$ or 3 the derivatives below exist, are continuous, and have the asymptotic properties indicated:

(3.8)
$$D^qA^{j+1}(\sigma,\tau) = \mathcal{O}(\sigma^{-q_1+(1-m)/2}) \qquad ((\sigma,\tau) \in \mathcal{S})$$
$$(j = 0,\dots,N ; \ 0 \le |q| \le 2(N-j)+2) ,$$

(3.9)
$$D^qB^{2k}(\sigma,\tau) = \mathcal{O}(\sigma^{-q_1-2k}) \qquad ((\sigma,\tau) \in \mathcal{S}) ,$$

$$0 \le |q| \le \begin{cases} N-2k & \text{if } N \text{ is even,} \\ N-2k+1 & \text{if } N \text{ is odd,} \end{cases}$$

$$1 \le 2k \le \begin{cases} N-2 & \text{if } N \text{ is even,} \\ N-1 & \text{if } N \text{ is odd.} \end{cases}$$

Note that $B^{2k+1} \equiv 0$. If we can establish (3.8) and (3.9), then $L(u - u_N)$ will behave so that the à priori estimate (1.8.9) of Theorem 8.1 of Chapter 1 can be applied, provided we are able to show that u_N satisfies the radiation condition, (1.3).

We shall use the following lemmas repeatedly in outlining the proof of (3.8) and (3.9).

Lemma 3.1. *If $a_1,\dots,a_s$ are given functions such that on $\mathcal{S}$*

$$D^{q'} a_j = \mathcal{O}(\sigma^{-m_j - q'_1}) \qquad (j = 1,\dots,s\ ;\ 0 \le |q'| \le |q|)\ ,$$

then on $\mathcal{S}$

$$D^q(\Pi_1^s a_j) = \mathcal{O}(\sigma^{-q_1 - \Sigma_1^s m_j})\ .$$

Lemma 3.2. *If $\gamma > 0$ and b is a function such that on $\mathcal{S}$*

$$1/b = \mathcal{O}(\sigma^{-\delta})\,(\delta > 0) \quad \text{and}$$

$$D^{q'} b = \mathcal{O}(\sigma^{\delta - q'_1}) \quad \text{for} \quad 0 \le |q'| \le |q|\ ,$$

then on $\mathcal{S}$

$$D^q(b^{-\gamma}) = \mathcal{O}(\sigma^{-\delta\gamma - q_1})\ .$$

Lemma 3.3. *If the vector functions v_i $(i = 1,2,3)$ have the property*

$$D^{q'} v_i = \mathcal{O}(\sigma^{-m_i - q'_1}) \qquad (i = 1,2,3\ ;\ 0 \le |q'| \le |q|)$$

on $\mathcal{S}$, then

$$D^q[v_1 \cdot (v_2 \times v_3)] = \mathcal{O}(\sigma^{-q_1 - \Sigma_1^3 m_i}) \quad \text{on } \mathcal{S}\ .$$

We omit the proofs of these lemmas as they are straightforward.

By Lemma 3.1 and the induction hypothesis, it is clear from (3.5) that to establish (3.8) and the smoothness of the A^j we need only to determine the smoothness and asymptotic behavior of A^0, B^{j-1}, and the coefficients of the A^j in (3.5).

Lemma 3.4. *If $m = 3$, n satisfies the hypotheses in Theorem 1.1, and $u_0 \in C^{2N+4}(\partial V)$, then for $0 \le |q| \le 2N+2$, the coefficients of the A^j in (3.5) (see (3.6)) have the smoothness and asymptotic properties listed: they each lie in $C^q(\mathcal{S})$ and*

$$D^q a = \mathcal{O}(\sigma^{-1-q_1})\ , \quad D^q c = \mathcal{O}(\sigma^{1-q_1})\ , \quad D^q d = \mathcal{O}(\sigma^{-q_1})\ ,$$

$$D^q f = \mathcal{O}(\sigma^{-1-q_1})\ , \quad D^q g = \mathcal{O}(\sigma^{-q_1})\ , \quad D^q d^{i,j} = \mathcal{O}(\sigma^{-1-q_1})\ ,$$

and

$$D^q e^i = \mathcal{O}(\sigma^{-1-q_1}) \qquad (i,j = 1,2)$$

on $\mathcal{S}$.

Sketch of Proof of Lemma 3.4. We illustrate the computation for a typical coefficient. We choose $d^{1,1}$ to be this typical coefficient. Since

$$d^{1,1} = \frac{iaH^{2,2}}{2}$$

to show that $D^q d^{1,1} = \Theta(\sigma^{-1-q_1})$, it is sufficient, by Lemmas 3.1 and 3.2, to show that

$$D^q a = \Theta(\sigma^{-1-q_1}) \text{ and } D^q H^{2,2} = \Theta(\sigma^{-q_1}) .$$

To show this we first prove $D^q J = \Theta(\sigma^{2-q_1})$. The desired behavior of J follows from that of X (given by Theorem 8.1 of Chapter 2) by Lemma 3.3 since $J = X_\sigma \cdot (X_{\tau^1} \times X_{\tau^2})$; see Theorem 6.1 of Chapter 2 for the asymptotic behavior of J. The desired behavior of $a = (nJ)^{-\frac{1}{2}}$ follows from Lemmas 3.1 and 3.2, our hypotheses on n in Theorem 1.1, and the properties of J which we have just established. It remains to consider $H^{2,2}$, defined in (3.4). The behavior of $D^q X_{\tau^2}$ is given by Theorem 8.1 of Chapter 2. We have determined the behavior of nJ. Therefore, by Lemmas 3.1 and 3.2 again, we obtain the desired behavior of $D^q H^{2,2}$. This completes our sketch of the proof of Lemma 3.4.

The function B^0 is $f(x)/n(x)$. By the hypotheses on f and n in Theorem 1.1, (0.45), formula (3.3) for the Laplacian, (3.6), Lemma 3.4, and Lemma 3.1, we easily establish (3.9) and the continuity of the derivatives involved in (3.9). We have thus outlined the proof of

Lemma 3.5. *If ∂V is convex relative to $n^{\frac{1}{2}}$ and n, u_0, and f satisfy the hypotheses of Theorem 1.1, then the desired smoothness and the growth properties of A^j and B^{2j} given by (3.8) and (3.9) hold. Further, ΔA^{N+1} and ΔB^k $(k = N-2, N-1)$ are continuous on $\bar{V}$ and are each*

$$\Theta([\sigma(x)]^{-(3+m)/2}) .$$

The order relations (3.8) and (3.9), which we have just established for $(\sigma,\tau) \in S_i$, do not vary with i (actually, $\tau = \tau^i$ on S_i)

$$\left.\begin{aligned} D^q A^{j+1}(x) &= \Theta([\sigma(x)]^{-q_1+(1-m)/2}) \\ D^q B^{2k}(x) &= \Theta([\sigma(x)]^{-q_1-2k}) \end{aligned}\right\} \quad (x \in \bar{V})$$

for the values of j and k used in (3.8) and (3.9). Note that

$$A^{j+1}(X(\sigma,\tau)) \equiv A^{j+1}(\sigma,\tau)$$

and

$$B^{2k}(X(\sigma,\tau)) \equiv B^{2k}(\sigma,\tau) \quad \text{for} \quad (\sigma,\tau) \in S_i .$$

Thus we conclude

Lemma 3.6. *Under the hypotheses of Theorem 1.1, if ∂V is convex relative to $n^{\frac{1}{2}}$, then*

$$|L(u-u_N)| = \Theta(\lambda^{-N}\sigma^{-2-(m-1)/2})$$

as $\lambda \to \infty$, *uniformly for* x *in* $\bar{V}$, $L(u - u_N)$ *satisfies the hypotheses made upon* g *in Theorem* 8.1 *of Chapter* 1, *and*

$$u(x,\lambda) - u_N(x,\lambda) \equiv 0 \quad \textit{for} \quad x \in \partial V .$$

These results are an immediate consequence of Lemma 3.5 and the definition of the approximate solution u_N.

4. The Radiation Condition

In order to apply Theorem 8.1 of Chapter 1 to $u - u_N$ and thus complete the proof of Theorem 1.1, it remains to verify that $u - u_N$ satisfies the radiation condition (1.3). To accomplish this it is sufficient to establish (2.1) for the functions (2.2). Recall that

$$\chi = \sigma - \sigma_3 .$$

Thus

$$\begin{aligned} \chi_r &= \chi_\sigma / (\partial r/\partial \sigma) , \\ &= 1(1 + \mathcal{O}(\sigma^{-1})) , \\ &= 1 + \mathcal{O}(\sigma^{-1}) . \end{aligned}$$

We have used the asymptotic behavior of X to perform this computation. It now follows from Lemma 3.5 and the proof in Chapter 2 that $\sigma(x) \sim r$ as $|x| \to \infty$ on V, that u_N does satisfy the radiation condition.

To complete the proof of Theorem 1.1 in the case that ∂V is convex relative to $n^{\frac{1}{2}}$ we need only apply the a priori pointwise estimate (1.8.9) to $u - u_N$.

5. General Obstacles

If ∂V is not by itself convex relative to $n^{\frac{1}{2}}$ but rather satisfies the hypotheses of Theorem 1.1 having the form (1.4), we construct the approximate solution u_N of (1.1) - (1.3) as follows. Let v_i be the approximate solution that can be constructed according to our Ansatz on the region spanned by the rays emanating normally from the patch S_i. Note that outside of the region spanned by these rays v_i is defined to be zero. At any point $x \in \bar{V}$, we now define

$$u_N(x) = \Sigma_i \, v_i(x) .$$

It is easy to see that the conclusions of Theorem 1.1 hold for this u_N. This point is more fully discussed in Chapter 0.

6. An Ansatz For More General Boundary Conditions

In this Section we consider exterior radiation-scattering problems of a more general type in which a linear combination of the solution and its normal derivative is prescribed on the boundary and, in addition, the coefficients in the boundary condition and the source terms depend upon λ. We suppose that $u(x,\lambda)$ is the solution of the following boundary-value problem (P*) :

$$Lu \equiv \Delta u + \lambda^2 n(x)u = f(x,\lambda) \qquad (x \in V) , \tag{6.1}$$

$$\Gamma u \equiv \alpha(x,\lambda)u + \beta(x,\lambda)\nu^* \cdot \nabla u = g(x,\lambda) \qquad (x \in \partial V) , \tag{6.2}$$

and

$$\lim_{R\to\infty} \int_{r=R} r\left|u_r - i\lambda u + \frac{(m-1)}{2r} u\right|^2 dS = 0 , \tag{6.3}$$

where $x \in R^m$ ($m = 2$ or 3) and ν^* is the unit exterior normal to ∂V. We assume, as previously, that ∂V is convex relative to $n^{\frac{1}{2}}$ or, alternatively that ∂V consists of a finite number of patches $S_1,\ldots,S_K$ that are locally convex relative to $n^{\frac{1}{2}}$, plus a portion S^* on which f and g are identically zero. To keep our hypotheses simple we assume that $n,f \in C^\infty(\overline{V})$, $n(x) \geq n_0 > 0$ on $\overline{V}$, and $\alpha,\beta,g \in C^\infty(\partial V)$. By keeping careful track of derivatives required these hypotheses could be relaxed. Further we assume that α, β, g, and f have the following asymptotic behavior as $\lambda \to \infty$, uniformly with respect to $x \in \overline{V}$:

$$\alpha(x,\lambda) = \lambda\alpha_1(x) + \alpha_0(x,\lambda) , \; D^p\alpha_0(x,\lambda) = \mathcal{O}(1) \qquad (0 \leq |p| < \infty , \lambda \to \infty) , \tag{6.4}$$

$$D^p\beta(x,\lambda) = \mathcal{O}(1) \qquad (0 \leq |p| < \infty , \lambda \to \infty) , \tag{6.5}$$

$$D^p g(x,\lambda) = \mathcal{O}(1) \qquad (0 \leq |p| < \infty , \lambda \to \infty) , \tag{6.6}$$

$$D^p f(x,\lambda) = \mathcal{O}(1) \qquad (0 \leq |p| < \infty , \lambda \to \infty) . \tag{6.7}$$

By following essentially the same steps described in Chapters 0 and 3, it is possible to construct an approximate solution of problem (P*) for $\lambda \gg 1$ and $x \in \overline{V}$. If $\beta(x,\lambda) \not\equiv 0$, the a priori estimates of Chapter 1 do not apply to problem (P*) ; and we therefore are unable to estimate the difference between the exact solution u and the approximate solution u_N^* of (P*) to prove that u_N^* is an asymptotic expansion of u as $\lambda \to \infty$, namely that

$$|u - u_N^*| = \mathcal{O}(\lambda^{-N+\frac{1}{2}(1+m)} r^{\frac{1}{2}(1-m)})$$

uniformly on $\overline{V}$. However, in the special case where $n \equiv 1$, $\alpha \equiv 0$, and $\beta \equiv 1$ and ∂V is convex, Morawetz [2] has obtained a priori estimates that should be applicable to yield this sort of result.

In the remainder of this Section we outline a rigorous argument for constructing an approximate solution $u_N^*(x,\lambda)$ with the following properties:

$$Lu_N^* = f(x,\lambda) + \mathcal{O}(\lambda^{-N} r^{-2-(m-1)/2}) \qquad (\lambda \to \infty) \tag{6.8}$$

uniformly on $\overline{V}$,

$$\alpha u_N^* + \beta\nu^* \cdot \nabla u_N^* = g(x,\lambda) + \mathcal{O}(\lambda^{-N}) \tag{6.9}$$

as $\lambda \to \infty$ uniformly on ∂V, and such that u_N^* satisfies the radiation condition (6.3).

Let S_i $(1 \leq i \leq K)$ be a typical locally convex patch of ∂V, and let T_i be the infinite tubular subregion of V covered by the ray field $\mathfrak{F}_i$ consisting of the optical paths emanating orthogonally from S_i. If $S_i = S_1 = \partial V$, then $T_i = V$, and we denote the field of rays covering $\overline{V}$ by $\tilde{\mathfrak{F}}$. We use an Ansatz of the type we used in Chapters 0 and 3; namely, if $x \in T_i$ we set

$$(6.10) \qquad \begin{aligned} u_N^*(x,\lambda) &= e^{i\lambda\chi(x)} \Sigma_{j=0}^{N+1} A^j(x,\lambda)\lambda^{-j} + \Sigma_{j=2}^{N+1} B^{j-2}(x,\lambda)\lambda^{-j} , \\ &\equiv e^{i\lambda\chi(x)} A(x,\lambda) + B(x,\lambda) . \end{aligned}$$

Note that we now are allowing the A^j and B^j to depend upon both x and λ (but in a bounded way). This has advantages in the sequel. Further note that although we assume that $x \in T_i$ in what follows, we have suppressed the dependence of A^j and B^j and u_N^* upon i.

We first determine B in (6.10) so that

$$LB = f + \mathcal{O}(\lambda^{-N}) .$$

We shall then determine A. Now

$$LB = nB^0 + nB^1\lambda^{-1} + \Sigma_2^{N-1} [\Delta B^{j-2} + nB^j]\lambda^{-j} + \Delta B^{N-2}\lambda^{-N} + \Delta B^{N-1}\lambda^{-N-1} .$$

Thus we choose:

$$(6.11) \qquad \begin{aligned} &\text{a)} \quad B^0 = f/n = \mathcal{O}(1) , \\ &\text{b)} \quad B^1 = 0 , \\ &\text{c)} \quad B^j = -\Delta B^{j-2}/n = \mathcal{O}(1) \qquad (j = 2,\ldots,N-1) . \end{aligned}$$

Then B is completely determined, and

$$(6.12) \qquad \begin{aligned} LB &= f(x,\lambda) + \Delta B^{N-2}\lambda^{-N} + \Delta B^{N-1}\lambda^{-N-1} \\ &= f(x,\lambda) + \mathcal{O}(\lambda^{-N}) . \end{aligned}$$

We now turn to the determination of A. We first consider the boundary condition (6.2). Simple computation shows that if $\chi = \sigma$, (σ,τ) being ray coordinates on T_i with $\sigma = 0$ on S_i, then on S_i

$$(6.13) \qquad [(\Gamma u_N^*) - g]/\Omega = A + \theta_1 \partial_n A + \theta_2 ,$$

where

$$(6.14) \qquad \begin{aligned} \Omega &= (\alpha + i\lambda n^{\frac{1}{2}}\beta) = \mathcal{O}(\lambda) , \quad \theta_1 = \beta/\Omega = \mathcal{O}(\lambda^{-1}) , \text{ and} \\ \theta_2 &= (\beta\partial_n B + \alpha B - g)/\Omega = \mathcal{O}(\lambda^{-1}) ; \end{aligned}$$

see (6.4) - (6.6) and note that $\nu^* = \nabla\sigma/|\nabla\sigma|$, which implies that $\nu^* \cdot \nabla\chi = |\nabla\sigma| = n^{\frac{1}{2}}$. We have assumed in (6.13) and (6.14) that $\Omega \neq 0$ on S_i. This will certainly be the case if

$$(6.15) \qquad \alpha_1 + in^{\frac{1}{2}} \lim_{\lambda\to\infty} \beta \neq 0 \quad (x \in S_i) , \quad \lim_{\lambda\to\infty} \beta(x,\lambda) \quad \text{exists,}$$

and λ is sufficiently large.

For $x \in S_i$, we impose the conditions:

(6.16)
$$\begin{aligned} &\text{a)} \quad A^0 \equiv 0 , \\ &\text{b)} \quad A^1 = -\lambda(\theta_1 \partial_n A^0 + \theta_2) = -\lambda\theta_2 = \mathcal{O}(1) , \\ &\text{c)} \quad A^m = -\lambda\theta_1\partial_n A^{m-1} = \mathcal{O}(1) \qquad (m = 1,\ldots,N+1) ; \end{aligned}$$

see (6.14). If we can determine functions A^j satisfying the conditions (6.16), then it will follow that

(6.17)
$$[(\Gamma u_N^*) - g]/\Omega = \theta_1\partial_n A^{N+1}\lambda^{-N-1} = \mathcal{O}(\lambda^{-N-2})$$
$$(N \geq 1) .$$

Therefore, on S_i

(6.18)
$$\Gamma u_N^* = g + \mathcal{O}(\lambda^{-N-1}) .$$

We next turn to the determination of A on T_i. We seek to choose A so that

(6.19)
$$L(e^{i\lambda X} A) = \mathcal{O}(\lambda^{-N}) \qquad (x \in T_i) .$$

Simple computation yields the result

(6.20)
$$L(e^{i\lambda X} A) = i\lambda[2\nabla X \cdot \nabla A + \Delta X A + (\Delta A)(i\lambda)^{-1}]e^{i\lambda X} ,$$

since

$$|\nabla X|^2 = n(x) ;$$

see Chapter 2. In order to satisfy (6.19) we proceed as in Chapter 3 to determine the A^j's by integrating the transport equations along optical paths emanating normally from S_i. To facilitate this we replace $\nabla X \cdot \nabla A$ by nA_σ in (6.20). The transport equations are:

(6.21)
$$\begin{aligned} &\text{a)} \quad 2nA_\sigma^0 + \Delta\sigma A^0 = 0 , \\ &\text{b)} \quad 2nA_\sigma^j + \Delta\sigma A^j = i\Delta_{\sigma,\tau}A^{j-1} \qquad (j = 1,\ldots,N+1) . \end{aligned}$$

where the Laplacian is computed in ray coordinates; see (3.3). We have used the result: $\nabla X \cdot \nabla A = |\nabla\sigma|^2 A_\sigma$ to obtain (6.21b).

We choose

(6.22)
$$A^0 \equiv 0 \qquad (x \in S_i \cup T_i) .$$

This choice is consistent with (6.16a). Now

$$\Delta X = n\partial_\sigma \ell n(nJ) ;$$

see (3.7). Thus (6.21b) becomes

(6.23)
$$A_\sigma^j + A^j\partial_\sigma \ell n(nJ)^{\frac{1}{2}} = \frac{i}{2}\Delta_{\sigma,\tau}A^{j-1} \qquad (j = 1,\ldots,N+1) .$$

Therefore

(6.24) $$A^1(\sigma,\tau) = \left(\frac{n(0,\tau)J(0,\tau)}{n(\sigma,\tau)J(\sigma,\tau)}\right)^{\frac{1}{2}} A^1(0,\tau) \ ,$$

where

$$A^1(0,\tau) = \lambda\theta_2 \ ,$$

with θ_2 given by (6.14). We rewrite (6.24) as

$$A^1(\sigma,\tau) = \mu(\sigma,\tau)A^1(0,\tau) \ .$$

Next we solve (6.21b) for A^j $(j \geq 2)$. The result is

(6.25a) $$A^j(\sigma,\tau) = A^j(0,\tau)\mu(\sigma,\tau) + \frac{i}{2}\mu(\sigma,\tau)\int_0^\sigma [\Delta_{s,\tau}A^{j-1}/\mu(s,\tau)]ds \ ,$$

where

(6.25b) $$A^j(0,\tau) = -\lambda\theta_1\partial_n A^{j-1} \ ,$$

with θ_1 given by (6.14). Note that $\partial_n A^{j-1} = n^{\frac{1}{2}}(0,\tau)A_\sigma^{j-1}(0,\tau)$.

This result should be compared with (3.5). The only major difference is in $\Gamma^j(\tau)$: If we integrated the Laplacian of A^{j-1} in (6.25a) by parts to remove σ derivatives, then we would obtain (3.5) with

$$\Gamma^j(\tau) = \{-cA_\sigma^{j-1} - gA^{j-1} - \lambda\theta_1 n^{\frac{1}{2}}a^{-1}A_\sigma^{j-1}\}_{(\sigma,\tau)=(0,\tau)} \ .$$

An induction argument analogous to the one followed in Section 3 would then lead to a conclusion analogous to Lemma 3.6; namely that on T_i,

$$L(e^{i\lambda\chi}A) = \Theta(\lambda^{-N}\sigma^{-2-(m-1)/2}) \ .$$

Moreover, one can also show as in Section 3 that

$$LB = f(x,\lambda) + \Theta(\lambda^{-N}\sigma^{-2-(m-1)/2}) \ .$$

Thus

$$L(u - u_N^*) = \Theta(\lambda^{-N}\sigma^{-2-(m-1)/2})$$

under hypotheses analogous to those of Theorem 1.1. Finally, an argument similar to that is used in Sections 3 and 4 would establish that u_N^* satisfies the radiation condition. Thus if one could establish the a priori estimates of Chapter 1 for the more general boundary condition (6.2), one could establish the analogue of Theorem 1.1 for the solution u of problem (P*) and the approximate solution u_N^* defined in this Section. This result, if proved, would generalize Ursell's result [3] to the three-dimensional case.

References for Chapter 3

1. C. S. Morawetz and D. Ludwig, An inequality for the reduced wave operator and the justification of geometrical optics. Comm. Pure Appl. Math. 21 (1968), 187-203.

2. C. S. Morawetz, Decay for solutions of the Dirichlet exterior problem for the wave equation, Comm. Pure Appl. Math. 38 (1975), 229-264.

3. F. Ursell, On the short-wave asymptotic theory of the wave equation $(\nabla^2 + k^2)\varphi = 0$, Proc. Camb. Philos. Soc. 53 (1957), 115-133.

CHAPTER 4
EXISTENCE OF SOLUTIONS

In this Chapter we prove existence of a solution to the radiation-scattering problem (P) of Chapters 1 and 3 in the case $u_0 \equiv 0$ and f is a function of λ as well as x. We assume that the data (f,n) and scattering obstacle ∂V satsify the hypotheses made in Chapter 1. We limit ourselves to the case of three dimensions $(m=3)$. Our existence theorem is dependent on the work of D. M. Eĭdus [1]. He studied more general equations than equation (3.1.3):

$$\partial(a_{ij}\partial u/\partial x^j)/\partial x^i + b_j \partial u/\partial x^j + (\lambda^2 + q(x))u = f , \tag{1.1}$$

where $a_{ij} = \delta_{ij}$ outside a compact set and the b_j have compact support. If $a_{ij} = \delta_{ij}$, $b_j = 0$, and $q = \lambda^2(n(x)-1)$, then (1.1) reduces to (3.1.3). We believe that our proof of Eĭdus's result in the special case of Problem (P) ((3.1.1)-(3.1.3)) has the advantage that it will generalize to operators of the form

$$\partial(a_{ij}\partial u/\partial x^j)/\partial x^i + b_j \partial u/\partial x^j + \lambda^2 n(x)u = f , \tag{1.2}$$

where $a_{ij} - \delta_{ij}$ and b_j do not have compact support. We also hope to apply the method used in this Chapter to study Problem (P) for piecewise smooth indices of refraction n and scattering obstacles with corners (piecewise smooth boundaries).

We call attention to related existence theorems by several authors. P. D. Lax and R. S. Phillips [2] studied Problem (P) in the case $\mathrm{supp}(n-1)$ compact in R^{2m}. Other related work, which we describe in terms of its specialization to Problem (P) is: research of J. R. Schulenberger and C. S. Wilcox [6], which applies to the case $\mathrm{supp}(n-1)$ compact and no scattering obstacle; a paper [3] of J. B. McLeod, which applies to the case where $|n(x)| \le Ce^{-p|x|}$; and a paper [5] by N. Myers and J. Serrin, which applies to the case

$$|x|^2(|n(x)| + |f(x)|) < \delta(|x|) ,$$

where

$$\int_2^\infty t^{-1}\delta(t)dt < \infty .$$

Our a priori estimates from Chapter 1 and Eĭdus's existence theorem [1] for the case $a_{ij} = \delta_{ij}$, $b_j = 0$, and q has compact support are the key to our existence proof. As is often the case with a priori bounds, if they imply uniqueness, then they imply existence as well. The idea of our proof is to consider a sequence of problems (P) with n replaced by n_j, where $\mathrm{supp}(n_j - 1)$ is compact and expands to all of V as $j \to \infty$, and to show that the sequence of solutions to these problems converges to a limit that is the desired solution of problem (P). Existence of solutions to the modified problems is provided by Eĭdus's results.

Let $\{n_j^{\frac{1}{2}}\}$ be an infinite sequence of indices of refraction such that

$$(1.3)\quad \begin{aligned} &\text{(i)} && n_j \in C^2(V) \cap C^1(\bar{V}) , \\ &\text{(ii)} && \operatorname{supp}(n_j - 1) \subset \{x : |x| < 2j\} , \\ &\text{(iii)} && n_j(x) \text{ lies between } n(x) \text{ and } 1 \text{ for } j \leq |x| \leq 2j , \\ &\text{(iv)} && \{n_j\} \text{ converges uniformly to } n \text{ on } V , \\ &\text{(v)} && n_j(x) = n(x) \text{ for } |x| \leq j . \end{aligned}$$

Let u_j be the unique solution of (3.1.1) - (3.1.3) with n_j replacing n. That such a solution exists was proved by Eĭdus. We now define

$$(1.4)\qquad v_{m,p} = u_{m+p} - u_m \qquad (p = 0,1,2,\dots)$$

and, for any function g defined on V,

$$(1.5)\qquad \|g\|_V^2 = \int_V |x|^{-2}|g|^2\,dx .$$

We shall prove that $\|v_{m,p}\|_V < \epsilon$ for any preassigned $\epsilon > 0$, as $m \to \infty$ for all $p \geq 0$.

Clearly,

$$(1.6)\qquad \begin{aligned} \Delta v_{m,p} + \lambda^2 n\, v_{m,p} &= \lambda^2(n - n_m)v_{m,p} - \lambda^2(n_{m+p} - n_m)u_{m+p} \\ &\stackrel{d}{=} F_{m,p} , \\ v_{m,p}(x) &= 0 \qquad (x \in \partial V) , \end{aligned}$$

and $v_{m,p}$ satisfies the radiation condition (3.1.3). For the moment we assume that

$$(1.7)\qquad \|r^2|F_{m,p}|\|_V < \infty \qquad (r = |x|) .$$

It then follows from the a priori estimates in Chapter 1 (Theorems 7.1 and 8.1) that there exists a $C > 0$, independent of $v_{m,p}$, λ, and x, such that for λ sufficiently large

$$(1.8)\qquad \max[(\max_{\bar{V}} |v_{m,p}(x,\lambda)|), \|v_{m,p}\|_V] \leq C\lambda^2\|r^2F_{m,p}\|_V .$$

Thus, using the definition of $F_{m,p}$, we find from (1.8) that for some $C > 0$

$$(1.9)\qquad \|v_{m,p}\|_V^2 \leq C\lambda^4[\|r^2(n - n_m)v_{m,p}\|_V^2 + \|r^2(n_{m+p} - n_m)u_{m+p}\|_V^2] .$$

Note that $\|v_{m,p}\|_V < \infty$ if $\|u_j\|_V < \infty$ for all j since $v_{m,p} = u_{m+p} - u_m$. Thus to show that

$$\|r^2F_{m,p}\|_V^2 < \infty$$

it suffices, because of our assumptions on the n_j and n, to prove that the V-norm of u_j is finite for all j, which we do below.

Next we write

$$(1.10)\qquad \Delta u_m + \lambda^2 n u_m = \lambda^2(n - n_m)u_m + f .$$

Our a priori estimate from Chapter 1, Theorem 7.1 yields, provided the V-norm of u_m is finite,

$$\|u_m\|_V^2 \le C\lambda^4 \{[\max_{r\ge r_1} r^4 (n - n_m)^2]\|u_m\|_V^2 + \|r^2 f\|_V^2\} \tag{1.11}$$

for each fixed r_1 and some $C > 0$, if $m > r_1$, since $n = n_j$ for $|x| \le j$.

To show that the V-norm of u_m is finite we write

$$\Delta u_m + \lambda^2 u_m = \lambda^2 (1 - n_m) u_m + f . \tag{1.12}$$

Let

$$H(x,x^0,\lambda) = \frac{1}{4\pi} \frac{e^{i\lambda|x-x^0|}}{|x - x^0|} , \tag{1.13}$$

and note that $|H| \in L_1(R)$ for any compact $R \subset V$ and $\|H\|_V < \infty$.

Then, using Green's second identity, we may write

$$\begin{aligned} u_m &= \lambda^2 \int_V [1 - n_m(x')] u_m(x') H(x,x',\lambda) dx' + \int_V H(x,x',\lambda) f(x',\lambda) dx' \\ &= \lambda^2 \int_{\{|x'|\le 2m\}\cap V} [1 - n_m(x')] u_m(x') H(x,x',\lambda) dx' + \int_V H(x,x',\lambda) f(x',\lambda) dx' . \end{aligned} \tag{1.14}$$

Therefore

$$\begin{aligned} |u_m(x,\lambda)| \le \lambda^2 [\sup_{\{|x|\le 2m\}\cap V} |u_m(x,\lambda)|][\sup_V |1 - n_m|] \int_{\{|x'|\le 2m\}\cap V} |H(x,x',\lambda)| dx' \\ + \|H(x,\cdot,\lambda)\|_V \|r^2 f\|_V . \end{aligned} \tag{1.15}$$

The quantity $\|r^2 f\|_V < \infty$ by hypothesis. It follows from the asymptotic behavior of H and (1.15) that there exists a $C > 0$ such that

$$|u_m(x,\lambda)| \le \frac{C}{1 + |x|} \quad \text{on } \bar{V} . \tag{1.16}$$

Therefore the V-norm of u_m is finite, and (1.8), (1.9), and (1.11) hold.

We now observe that if $m > r_1$ and λ is fixed,

$$C\lambda^4 \max_{r\ge r_1} [r^4 (n - n_m)^2] < \tfrac{1}{2} , \tag{1.17}$$

provided r_1 is large enough. This follows from the far-field behavior of n which we have assumed. Therefore, by (1.11),

$$\|u_m\|_V^2 \le 2C\lambda^4 \|r^2 f\|_V^2 , \tag{1.18}$$

so that the V-norm of u_m is uniformly bounded in m for λ fixed.

If we use the uniform bound (1.18) in (1.9), we easily show that for λ and m sufficiently large, λ fixed, and all $p \ge 0$

$$\|v_{m,p}\|_V < \epsilon \tag{1.19}$$

for any preassigned $\epsilon > 0$. Therefore $\{u_j\}$ converges in the V-norm, say to u;

and hence almost everywhere on V

$$(1.20) \qquad u(x,\lambda) = \lambda^2 \int_V [1-n(x')]G(x',x)u(x')dx' + \int_V G(x',x)f(x',\lambda)dx' ,$$

$$u = 0 \quad \text{on} \quad \partial V ,$$

where G satisfies the radiation condition, $G(x',x) = 0$ for $x' \in \partial V$, and

$$(1.21) \qquad \Delta_{x'} G + \lambda^2 G = \delta(x'-x) \qquad (x,x' \in V) .$$

We note that G is square integrable in the V-norm and

$$(1.22) \qquad r'^{-(2+\epsilon)} \partial G/\partial x_i' \qquad (i = 1,2,3)$$

lies in $L_1(V)$ if $\epsilon > 0$ [4].

The integral equation (1.20) shows that u is continuous; and since the quantities in (1.22) lie in $L_1(V)$, u_{xi} exists and is continuous. One can now show by standard argument that u satisfies (3.1.1). To show that u satisfies the radiation condition one uses (1.20) and the fact that G satisfies the radiation condition. This completes the proof that at least one solution of Problem (P) exists. Uniqueness follows directly from the a priori estimates of Chapter 1. We have thus proved:

Theorem 1.1 *If* $u_0 = 0$, f *is a function of* λ, *and* $(n,f,\partial V)$ *satisfy the hypotheses of Theorems* 7.1 *and* 8.1 *of Chapter* 1, *then the Problem* (3.1.1) - (3.1.3) *has one, and only one, solution.*

To extend this theorem to nonzero boundary data is not hard. For example, the method just used to prove Theorem 1.1 can be used. One can easily manage the boundary data contributions that will appear in (1.11), (1.14), and (1.15). Another, less satisfactory approach, but quicker, is to extend the boundary data u_0 to a function of compact support on $\overline{V}$. If u_0 is smooth, then its extension u_0 will be smooth also, and we can write $u = v + u_0$. By Theorem 1.1 there exists a solution v of

$$\Delta v + \lambda^2 n v = -(\Delta u_0 + \lambda^2 n u_0) + f$$

$$v|_{\partial V} = 0$$

$$\lim_{R\to\infty} \int_{r=R} r|\mathcal{D}_1 v|^2 = 0$$

Then $u = v + u_0$ is a solution of (3.1.1) - (3.1.3).

THE END

References for Chapter 4

1. D. M. Eĭdus, *Some boundary-value problems in infinite regions*, Izv. Akad. Nauk SSSR. Ser. Mat. 27 (1963), 1055-1080. = A.M.S. Transl. (2) 53 (1966), 139-168.
2. P. D. Lax and R. S. Phillips, *Scattering Theory for the Acoustic Equation in an Even Number of Space Dimensions*, Indiana U. Math. J. 22 (1972), 101-134.
3. J. B. McLeod, *The Analytic Continuation of the Green's Function Associated with Obstacle Scattering*, Quart. J. Math. Oxford, Ser. (2), 18 (1967), 169-180.

4. C. S. Morawetz and D. Ludwig, An inequality for the reduced wave operator and the justification of geometrical optics, Comm. Pure Appl. Math. 21 (1968), 187-203.
5. N. Myers and J. Serrin, The Exterior Dirichlet Problem for Second Order Elliptic Partial Differential Equations, J. Math. and Mech. 9 (1960), 513-538.
6. J. R. Schulenberger and C. S. Wilcox, The Limiting Absorbtion Principle and Spectral Theory for Steady-State Wave Propagation in Inhomeogeneous Anisotropic Media, Archive for Rat. Mech. and Anal. 41 (1971), 46-65.

INDEX